A Team Sergeant's Handbook

A Team Sergeant's Handbook

By Thomas Kelly

Edited by CSM Kevin Dorsh

A Team Sergeant's Handbook

by Thomas Kelly

ISBN 978-1-956904-11-6

Printed in the United States of America

Published by Blacksmith LLC

Fayetteville, North Carolina

www.BlacksmithPublishing.com

Direct inquiries and/or orders to the above web address.

For my wife Donna.

Contents

Foreword

I'll start this foreword by saying my Team Sergeant time was from 2004 to 2007. Although that was a relatively small part of my 32 years of service, it will always stand out as the most memorable. A majority of my experience and lessons learned during those dynamic years continued to be relevant to my last assignment as the Command Senior Enlisted Leader at SOCCENT.

I first met Tom, then SSG Kelly in 2005 prior to our unit's rotation to Iraq. He was on a sister Operational Detachment–Alpha (ODA) that was led by a friend of mine, so we had an opportunity to train together on numerous occasions. Even though Tom was a few years out from his time as a Team Sergeant, I observed him to be a natural leader among his peers. As the years – and deployments – went by, I was able to keep tabs on Tom and watch him mature into a seasoned Green Beret.

Early on in this book, Tom will discuss his transition from team bro to Team Sergeant. He lays out the difficulties he faced navigating this new terrain, and the lessons learned from each. Unfortunately, being bro'd up, and then put into the position of being the boss – on a team you are currently assigned - is difficult. This is the core reason why, when guys make Master Sergeant, they are moved down the hall, to a different Company, or to another Battalion altogether. We all know what that means though, he was a great teammate! He was trusted with all the secrets of the team and its members.

The information laid out in this handbook is universal and can serve all operational Groups, Battalions, Companies, and ODAs. The principles outlined are timeless and have been proven to be effective in both garrison and combat environments. These enduring principles serve as a solid reference to the many First-year questions every new Team Sergeant has. Although everyone's experience will vary - due to many things that cannot be controlled - this guide is a goldmine for those: taking a team soon, newly assigned, or even those halfway through their time in the seat.

As I read the book, I noticed input from multiple Senior Non-Commissioned Officers from different Operational Groups, and from slightly different times in our history. Every topic discussed shared a commonality that most Team Sergeants face but highlights how the finite details were of course different. Tom does a fantastic job using his, and others' stories to provide context to the leadership principles used in this book. The storytelling method, I find, has always helped me remember and apply important concepts.

At the end of the day, Tom was a successful Team Sergeant. By writing this book, he embodied the ARSOF attributes, Courage, and Personal Responsibility. Courage: to the best of my knowledge, Tom is the only guy to write a book on this topic. He laid his experience – good and bad – out there to be judged by the community, and to pass the torch for generations to come. Personal Responsibility: owning his mistakes. More importantly, he goes on to provide you with alternative ways to approach those situations.

Never forget that you are about to move into the best position in all the Special Forces career field!! Hopefully, the Battalion

Foreword

Command Sergeant Major has assigned you to an ODA that is aligned with your specific attributes, and special skills. As with everything in the Army though, at the end of the day, you will move out and lead whatever, and wherever you are LUCKY enough to have the opportunity.

You will be entrusted with our most valuable commodity, Green Berets. Absolutely use this book to help you navigate your time in the seat. As you prepare to take on the best job in the universe, use this book to get your headspace right for the overwhelming amount of responsibility you are about to assume. Solicit feedback from your leadership, your peers, and from your subordinates. All of the above will give you a reliable azimuth check.

I am envious of your upcoming journey. Remember to trust your instincts, train hard and most importantly, have fun while doing it!

Good luck, and De Oppresso Liber!

Sincerely,

Robert Flournoy CSM(Ret.)

Letter from the Editor

Plain and simple - I wish I would have had a chance to read this book before I went into position as a Team Sergeant.

The Team Sergeant position for our Special Forces Operational Detachment – Alphas and each of our manuever elements within the Special Forces Regiment is the **Center of Gravity** for all that we do. It is not the Officers at echelon above the Detachment, or even the Detachment Commander as some movies like to portray – it is the Team Sergeant. The Team Sergeant is the Senior NCO on the Detachment who is responsible for each and everything that Detachment does or fails to do. They are the fuel that makes the machine run. From serving as the Master Trainer, the mentor to the Detachment Commander, to the leadership guru who sets the conditions to allow the Detachment to function seamlessly in situations of extreme stress. It is the Team Sergeant who is the Center of Gravity for the Special Forces Regiment.

In years past, how did we "annoint" our Team Sergeants? How did we select a person for such a critical position in our organization? Simply, it was the CO SGMs and BN CSMs that nominated someone, tapped them on the shoulder and let them know – "you are going in." No development, no training, and until now, there was no manual that could at least point you in the right direction and provide you a framework on how to at least approach preparing for this momentous job. Tom has provided our

Regiment with exactly that – *A Team Sergeant's Handbook*.

Over the last twenty years, the role of the Team Sergeant has become even more critical than it had been in years past. As our force becomes "younger" in Army experience and the hard lessons learned from combat evaporate, it is the Team Sergeant that must continue training and developing our Soldiers within the Regiment for the future fight. *A Team Sergeant's Handbook* provides you the basics, it provides you an overview of all the areas that you must think about and consider if you are going to do the job well. It provides you keen insight on the importance of understanding who you are as a leader and points you to phenominal resources to help you with that journey. Until now, we have had nothing like this – a single resource, consolidated in a single book that held the keys to success, the answers to the test on how to be a successful Team Sergeant.

A Team Sergeant's journey from the beginning when he takes the helm to the end of his 24 or 36 month tour is a story of growth and maturity of a leader; typically with a ton of lessons learned and simply A LOT of mistakes. Trials and tribulations normally mark the beginning of the tour while towards the end, the Detachment is running smooth, running on all cylinders without missing a beat. That is not hapinstance. That is due to the experience gained through the Team Sergeant. *A Team Sergeant's Handbook* provides the tools to flaten that curve of exploration, the trials and tribulations, and allows the chance to confidently hit the ground running, starting day 1 in the position.

What I enjoy most about this book is the fact that it is the culmination of one Team Sergeant's journey, Tom's journey. A culmination of multiple years in the position as a Team Sergeant coupled with a serious deep reflection years later that plainly pointed out some areas where he did not perform as well as he would have wanted. It is a body of work that has taken over 20 years - 20 years of Army experience, and multiple years to reflect and write. It is that type of vulnerablity that will allow the next generation of leaders to hopefully learn from his "pain" and not have to experience the same. That old addage – we learn through pain – hopefully can be avoided by reading this book and heeding the advice that Tom simply lays out with ease within these pages.

I hope you enjoy this book as much as I did. For those of you preparing to take the helm of the best position within the Army, the Team Sergeant – good luck! Enjoy it – the time goes fast. But rest assured, you made a great choice by reading this book and are on azimuth for a successful tour as a Team Sergeant.

De Oppressor Liber
Nous Defions

Kevin Dorsh
10th SFG(A) CSM
Editor *A Team Sergeant's Handbook*

Preface

If you thought a living legend within our Regiment wrote this book, sorry to disappoint. When my Team Sergeant time began, I looked at my peers and at those who had come before me... humbled hardly explains my state of mind. Even though it was a tremendous honor to be entrusted with the position, I immediately knew I was now competing in the big leagues. When my time in the seat was done, I conducted (over many years) a serious self-reflection. I found that although I did many things right, I also fell short in numerous areas. This book is the sum of that deep dive.

As a Regiment we have done ourselves a huge disservice by not having a starting point for new Team Sergeants. We put a lot of emphasis on one's ability to quickly adapt to a situation and come out on top. Sure, it can be argued that is the nature of our profession. Congruently, we should all agree, a Team Sergeant can make or break the team. I believe it is one of, if not the most critical position within our formation. Why then do we expect aspiring Team Sergeants to jump into the most essential and dynamic role within SF, without so much as a compass heading? While I hold no naive expectation to have a silver bullet remedy, this book tackles what I believe to be a good starting point. As such, this volume is designed to discuss the critical tasks, skills, and knowledge every aspiring, and current Team Sergeant should have at his disposal.

Looking back at what my Non-Commissioned Officers Education System (NCOES) path was comprised of, and compare it to what is available today, I am envious. The addition of the Master Leader Course (MLC), in my opinion, has been a fantastic step in the right direction. If I had the exposure to the following information, I truly believe my time in the seat could have played out differently.

Extracted from the MLC's ARSOF Senior NCO Fundamentals POI[1]:

> Purpose: To educate Senior NCOs on essential Army Special Operations Forces (ARSOF) topics designed to supplement the Master Leader Course curriculum, and to ensure that ARSOF Senior NCOs are provided with the requisite knowledge, competencies, and leader skills to effectively perform the wide array of duties and responsibilities required of an ARSOF Master Sergeant and First Sergeant.
>
> Phase Scope: The course is designed to challenge and educate selected ARSOF Sergeants First Class promotable (SFC-P) personnel in the areas of SOF history, professional writing, communication skills, public speaking, Joint planning and Joint planning systems, critical thinking, operating in JIIM environments, lessons learned process, TSOC roles and responsibilities, statutory and funding authorities, legal issues and processes, identity management, G8 (force management) staff, and the preservation of the force and family.

---

[1] ARSOF Senior NCO Fundamentals POI, course: 1-250-C6 (ARSOF), Approved OCT 1 2019, Version 01.0 .

Although the above is a lot to cram into a 40-hour training week, it gives the up-and-coming Team Sergeant a peek behind the proverbial curtain. Depending on his previous leadership this might be his first look at the big picture, and how SF fits into the national strategy. Additionally, it gives him some tools in which to navigate it.

The Team Sergeant Assessment Program (TSAP) is another recent development confronting the above issues. Currently, each of the operational Groups have implemented this program. For 10th Group, the primary objective of TSAP, contrary to the title, is to provide mentorship for newly promoted E8s, and guys with a solid OML. As with any program, the POI will change overtime to adapt to our leadership challenges, for now here is a quick snapshot of what to expect.

When the course begins each future Team Sergeant is assigned a mentor from another Company or Battalion (post Team Sergeant, SGM, etc.) to assist him in navigating the POI. The mentors are cross pollinated for good reason. As explained to me this method equips the future Team Sergeant with a mentor that is out of his Support Channel. This provides someone to engage with to seek guidance on small or large issues, who is not deeply invested in your Company.

The course is run roughly three times per year, one week in length, and starts with an ACFT. A good portion of the week is spent on Unit Training Management (UTM). They will utilize Army Manuals (FM, ATP, ATTPs etc.) and classroom instruction to gain a better understanding of how to develop a comprehensive training plan for their ODA. To close out the training week, the cadre have mystery events to test the aspiring Team Sergeant's abilities against common, and complex scenarios that ODAs face across the Regiment.

As every man who has ever served as a Team Sergeant can attest, regardless of what MLC, TSAP, and even this book can provide, the largest influence of future Team Sergeants, are the guys currently in the seat. Every second served in the position is under constant observation: physical fitness, professional bearing, competency, resiliency, morality, and ethics are cataloged (consciously or subconsciously) by our guys. These factors, and many others will shape how future leaders will handle their time running an ODA.

The following guide is comprised of things I succeeded and failed to do during my time. Chapter 1 is comprised of some hard lessons learned, and a few general rules to help you steer clear of the landmines I stepped on. Chapter 2, the often unsung and unappreciated aspects of a Team Sergeant's duties - administration and counseling. Then, Chapters 3 through 5 will survey the domains of training and leadership. Chapter 6 details what I believe should comprise a Team Sergeant's library. Chapter 7 holds the remaining professional tips I mustered up from my time, but could not fit into the other chapters. Finally, Chapter 8 concludes with a checklist that will aid a new Team Sergeant take the reins of his team.

As one makes their way through this book, they will notice I cited a few authors, and Sergeants Major numerous times. If you have been in the Regiment for more than a few years, you know that everyone has an opinion on everything, and everyone. One day you hear how awesome someone is for reasons x, y, z, and the next you hear they are a huge tool. Regardless of this, put the bias aside for the time required to read this book. At the end of the day, these authors took the time to write down their experiences for us to learn from.

Another phrase you will see often in this book is: "Don't recreate the wheel." I wasted a lot of time early on during my time in the seat trying to figure things out on my own. A quick conversation with a peer, or daily effort reading leadership books, and military manuals would have saved valuable time. Put aside your ego early and show up humble, you do not have all the answers walking into this job. By the time you have it figured out, you'll be exiting the Team Room for good.

The tasks of an Operational Detachment Alpha (ODA) will invariably change, Areas of Responsibilities (AORs) will shift, and world events will drive training in all directions, but the basics will always win the day. The information in this book may be from an old guy, but the duties and responsibilities of the Team Sergeant will always be centered on the information in this guide.

The hope is you read this prior to being assigned a Team. If your current Team Sergeant is taking the time to mentor you and others to do the job, use the following as a checklist. If not, now you have the time to prepare yourself for one of the most difficult yet rewarding jobs in the Regiment.

Otto von Bismarck said:

> "Fools say that they learn by experience. I prefer to profit by other's experience."

Writing this guide probably makes me the fool, but I hope the following information helps during your time as a Team Sergeant.

Dedicated adherence to spartan standards, maintenance of continuous operational readiness for the conduct of multiple, complex missions; complete faith in the capability to execute those missions and the willing acceptance of hazards far beyond the normal call of duty have made the 10th Special Forces Group the unit with the greatest combat potential in the Armed Forces.

- Col. Aaron Bank, quote from the 1956 10th Special Forces yearbook

1

General Rules of the Road

1-1 Welcome to the big show! Being a Team Sergeant is the best and worst job in the Army. In fact, I can say without hesitation, at times you must be prepared for adult daycare activities.

Being a Team Sergeant involves managing personalities and employing (task organizing) the right guys with the right teammate, on the right mission, or objective.

In that vein, three components crucial to a Team Sergeant are <u>Organization</u>, <u>Caring</u> and <u>Ownership</u>. If you're *organized*, then the administrative distracters will not interfere with your training. If you *care*, then you'll take the steps necessary, choosing the hard right over the easy wrong, and go the extra mile to ensure your ODA[1] is fully prepared to execute its mission. *Ownership*, to me means branding the ODA. Make the effort to develop a culture you want the team to have. Or if the team has a respected brand, work hard to maintain it. Ownership also requires you to accept responsibility for everything the team does or fails to do.

---

[1] ODA-Operational Detachment-Alpha. Understanding my target audience, I did not define each acronym throughout the book. See the Glossary for a complete list.

As this book progresses, I have tied in my lessons learned with the 12 Principles of Leadership. Take the time to review the principles located in Appendix II before you continue. I borrowed these principles from Paul Lefavor's *Tactical Leadership* and *US Army Small Unit Tactics Handbook*. That being said, here is a list of General Rules to follow, some pitfalls to avoid, a few dos and don'ts, and my experiences with them all.

1-1.1 **Pitfalls.** A pitfall many Team Sergeants have fallen into is trying to be everyone's friend, or better said, attempting to achieve **Popularity**. If you are concerned about the boys liking you, then you are not focused on the mission, or the men's best interest.

1-1.2 The polar opposite of this issue is the guy who says, "I am not here to make friends." The statement is not wrong; the level of detachment required to execute the job does limit how close you can get with the guys. But, when announced to the team it can be understood to mean something else, such as:

- The team is only a steppingstone
- Zero human connection capability
- 100% detached
- Zero leniency for all infractions

1-1.3 Directly linked to the pitfall of popularity is, you are no longer **one of the boys**. It can be a rough transition from a team bro to Team Sergeant. You must elevate to the adult in the room. The trend of selecting younger E-7s for promotion, which started around 2010, seems to be the new norm in the Regiment. Hopefully these young NCOs are being moved down the hallway at a minimum.

1-1.4 **Assuming responsibility of your current team.** If you are selected to take over your current Team, be sure to separate yourself from the boys. This may seem counterintuitive to everything you know to be right, in fact, it will feel like a punch to the gut. The job comes with an awesome amount of responsibility, and the separation or better said, the **detachment,** gives you a chance to see the big picture. Your guys will understand the new separation, as long as you give them the WHY behind your actions and remain consistent with those behaviors.

1-1.5 **The big picture is the welfare of the WHOLE team and the mission**. Without detachment, the possibility of siding with a buddy over the general welfare of the ODA is real. Of course, there is a dichotomy here, total separation, or "Distancing yourself from those you lead" is not an option. Kyle Lamb warns against doing this in his book *Leadership in the Shadows*. This is just one of the of things to GUARD AGAINST as a military leader–

> Distancing yourself from those you lead. Camaraderie is extremely important in the military. You will be living and working with these people for extended periods of time.[2]

[2] Lamb, K. Ch. 9 Understanding Military Leadership. In *Leadership in the Shadows*.

Knowing as much as possible about the men is critical. Without this knowledge, you will be hard-pressed to manage personalities, and employ the right guys with the right teammate, on the right mission and/or objective. Every decision you make has effects on the team's mission and overall well-being - No pressure!

1-2. Fun fact, you cannot make everyone happy. A Team Sergeant must make decisions such as: Top rated NCO, task organization (for all mission sets), who gets which school, who goes on ADVON, SWCS assignments, etc. These decisions MUST be based on what is best for the whole Team, the individual's professional development, and mission success.

1-3. So, how does a leader influence the team to execute difficult tasks? Which leader attributes and actions build-up, and sustain Leadership currency?

1-3.1 While trying to find the answer to the above questions, I Googled my way to franklincovey.com. Here I found Stephen M. R. Covey's "The 4 Cores of Credibility"[3], which are part of his book *Leading at the Speed of Trust*. I believe he nailed it by anchoring his 4 Cores in Trust.

Covey states:

> Trust is a function of two things: character and competence. Character includes your integrity,

---

[3] Steven M.R. Covey, The 4 Cores of Credibility, resources.franklincovey.com

> your motive, your intent with people. Competence includes your capabilities, your skills, your results, and your track record. The good news is that we can increase our credibility, and we can increase it fast, particularly if we understand the four key elements, or four Cores that are fundamental. Two of these cores deals with character; two with competence.[4]

To break that paragraph down further, for character the following "Cores" are required:

- Integrity
- Intent

And for competence:

- Capabilities
- Results

1-3.2 If a leader weighs every action and decision against these four "Cores," their leadership currency will continue to grow. When any one of these is not factored into your actions, this is when your reputation begins to decline. We will revisit leadership currency again in Chapter 5.

1-3.3 Why would I quote a civilian within the first chapter of the book? Well, when conducting an honest self-assessment, one must be willing to let the research go outside one's comfort zone, especially within the art of

[4] Steven M.R. Covey, The 4 Cores of Credibility, resources.franklincovey.com

leadership. Many of my lightbulb moments have come while reading or listening to leaders who are not Green Berets, have not served in the Army, or in any military capacity whatsoever.

1-3.4 **Lesson learned**, I failed to fully make the leadership mindset transition early on. I initially tried to use my status as "one of the boys" and focused my energy on popularity. Had I worked on acquiring Leadership Currency, and applied the 12 Principles of Leadership, many of my failures could have been avoided. This approach, as proven through many books written about successful military leaders, would have properly earned the trust of the men.

1-4. Comfort zone, or better said by CSM Dorsh:

> "Put yourself in uncomfortable situations. It makes you sharper to be nervous."

Working outside your comfort zone can be a daunting task, especially when the pressure put on Team Sergeants is no joke. From day one, until you turn in your team gear, you are responsible for, but not limited to:

- The professional lives of nine NCOs
    - Master trainer, coach, mentor, and manager
- Reverse mentorship of the Captain
- Preparation, execution, and completion of the ODA's assigned mission (you and the Captain share this Task)

- SOP development, and implementation
- Specialty infill capability
- Retention
- Every administrative action

It is understandable with the above pressure, guys will stick with what they know when developing training.

1-4.1 In order to grow as a team (and as a leader), you need to exit your comfort zone. Leaving the comfort zone does not mean jumping headfirst into Full Mission Profile (FMP) training events, with super complex scenarios involving a robust roll player package, EOD, dog teams, etc., before the team is ready.

Your training approach should be linear in nature (Individual, Collective, FMP – covered in Chapter 3), and your focus throughout should primarily be on the basics. If done correctly the Collective phase will advance with a natural progression of difficulty, longevity, and complexity. Here is where you begin to introduce scenarios that push you and the team out of the comfort zone. Ensure the team, and the leadership are pushed hard, both physically and mentally. Lastly your FMP needs to have a realistic feel and test your team on what is projected to be the most difficult part of the upcoming mission.

1-5. Success is achieved through failure. I could not help myself; this training fact will show up multiple times throughout the book. Success is built on getting it wrong, dissecting what went wrong, and pushing past assumed limitations. This fact includes you. Although you have put in the mileage to be where you are, know what you know, you will learn new things. When you push past your known limitations, it is a good chance that the guys who were apprehensive before, will follow your lead.

1-6. Lead from the front. Definitely not a new concept for this audience. As mentioned, a few times throughout the book, all eyes are on you now. If you are not up front, leading the charge, doing the hard work with the team and getting your hands dirty, someone will step up and fill the leadership gap. People in our ranks have issues following a guy that won't do the work with them. Yes, this job has requirements which can take you away (training meetings, briefings, etc.), do what you can to be with the team when the work gets tough.

1-7. Not knowing everything is ok. Team Sergeants in the ranks now are required to be proficient or knowledgeable on all tasks from the basics to electronic warfare. You are not a SME in every SF task, and it's okay to admit that to your guys. Admitting you do not know something takes courage in our ranks, especially if you have been in long enough to be an E8. Your days of buying beer in bulk for the team fridge should be well behind you. Don't let pride get in the way, the boys will appreciate you pushing trust down to them when it comes to training. Change will happen while you are in the seat, whether it is technological or cultural in nature, be ready to lean on your smart guys to help your team implement and/or adopt it.

1-7.1 Even if you happen to know the block of instruction, step out of your comfort zone and allow some of your younger guys to take the lead. Allowing your guys to take lead on something falls under the ***9th Principle of Leadership, Develop a sense of responsibility in your subordinates.***

1-7.2 If you hit a wall during planning or have no idea where to start on a project, engage other Team Sergeants in the Company, Ops Sergeants at Battalion, or Group. Ask how they have handled similar situations and planned training. Read through other ODA's training concepts for inspiration, doing so can let you know if your initial direction was correct, missing the mark, or if you are trying too hard. Once you have an idea, or a vision how the training event should go down, use everyone on the Team (when time permits) to develop plans.

1-8. When time is NOT a luxury, ***Leadership Principle #6: Make sound and timely decisions*** comes into play. You will face and make many decisions during your tenure. Some decisions are simple and require no thought at all. Others require time to research a regulation or speak with a subject matter expert.

Then there are those decisions that you must make right now. When you have no time to dig into the pros and cons, review data points, or discuss options with your peers, and subordinates. Do not be afraid to make these calls, sure it cuts deep when you are wrong, but not making a call in our line of work is far worse.

1-8.1 Reaching back to the Leadership currency paragraph, not making a decision is a quick way to lose it. Looking forward in this chapter - **No one is perfect, especially you** - for those times you make a bad call. Bottom line, you are paid the "big bucks" to make decisions. The men, including the Captain, will always look to you during difficult times. Use your experience, execute a rapid assessment, and make a call and/or recommendation. Not making a decision is a decision not to decide.

1-9. Don't show favoritism. Although this should be common sense, and never need to be written down, it's easy to click with a similar-minded individual. Favoritism can be displayed in a few ways. One example is spending more time with one, or a few guys on the Team. Whether this is during non-duty hours or training, it can be viewed poorly. This is especially important if the guy(s) you click with get something which is perceived as a leg up; advanced school slots, a higher enumeration on NCOERs, etc.

1-10. *Leadership Principle 2: Know yourself and seek self-improvement.* The pitfalls mentioned above, and later in this chapter are the ones I felt that met the scope of this book. There are countless more, but I will not air all of my dirty laundry. Good news though, whether you self-identified an issue, or someone did the right thing and confronted you on a problem, you can rectify it.

1-10.1 If you find yourself flirting with one or more of the above (or later) pitfalls, seek self-improvement. Don't know where to start? Engage with a peer or hit up the SGM. Ask for their approach to your situation, be open to what you hear, and apply what they say if it makes sense. If you want to dig deeper, there are plenty of options, but to help slim it down I found Paul Lefavor's *Tactical Leadership* to be extremely helpful.

Paul's first chapter is titled "The Philosophy of Leadership." It covers what I believe to be the full spectrum of topics important to the art of leadership. This chapter redefined what I thought I knew about this critical topic.

1-10.2 Specific to this chapter though, skip to page 34 in *Tactical Leadership*. Here, Paul has a fantastic acronym that cuts straight into character flaws. He uses the acronym BLEMISH to help us easily remember them.[5,6]

B – Blame shift
L – Lackey
E – Egomaniac
M – Micromanager
I – Inability to keep your cool

[5] Paul LeFavor, *Tactical Leadership* (Fayetteville, NC: Blacksmith Publishing, 2017), 34. Also covered on the Pinelander Podcast, Episode 15, Tactical Leadership, March 18, 2022.

[6] See Appendix III for the full description of Paul LeFavor's acronym.

S – Self-control
H – Hypocrisy

As you read Paul's acronym BLEMISH, think back to your past leaders, did they embrace one or more of these traits? How did your performance differ, or suffer when serving under these conditions? Now look at yourself, are you dabbling in these flaws?

Later in Chapter 5, On Leadership, I will cover my additional takeaways from Paul's book *Tactical Leadership*.

1-10.3 Obviously Paul does not own a monopoly in the Leadership market. There are more than a few Leadership books out there with wonderfully thought-out principles available. In fact, I provided two additional lists in Appendix II, page 274.

1-10.4 Recently I was given a new (2020) leadership book. Author, and former leader in the SOF community Colin Greata, has a chapter in his book *Always Endeavor* called "Leader Approaches." I found his view on leadership principles enlightening because he prepares the reader to "Let go of the handrails of should and should not."

Yes, he does have a list of Maxims to review and apply, but I will let his chapter introduction speak for itself:

> In this chapter, we will codify some practical leader approaches. These may be what you

consider leadership principles, but even these principles are not laws: leaders at the master's or doctoral level of leadership will see these as general guidelines, not hard and inviolable rules. These principles will *characterize* a leader's general conduct, but *not define* every single behavior.[7]

1-11. Getting Read On. You did all the work to obtain a TS clearance, go use it. Your new job takes on a lot of responsibilities, one being – ensure the guys take EVERY mission seriously. The best way, in my opinion, to fully understand how our Team's mission is tied into the assigned AOR is done by getting a read on completed.

1-11.1 Take the time needed to figure out the strategic importance of your mission. To do this, read and understand your Battalion's Annual Training Guidance, then dig deeper. Get ahold of 1st SFC(A), USASOC, and USSOCOM's Annual Training Guidance and vision statements. Between the different Annual Training guidance documents, and getting read on, you will gain a full understanding of the strategic reasons behind the places we visit. Only then will you grasp how important that mission, that Partner Force, or that host nation relationship is. This is what you should be relaying to the guys. No matter how "lame" you think the assigned

---

[7] Always Endeavor, A Developmental Guide for In Extremis Leaders, Colin Greata, Part III, pg. 194, Leader Approaches.

mission is, you should refrain from voicing a negative opinion.

1-12 Assessment Period, *Leadership Principle 11: Build a Team.* Use your first month to get to know the Team. Doing this can prevent trying to change things before you understand the culture of the ODA. During this time frame, you should observe and take copious notes to answer probing questions like, but not limited to:

- What does the next six-to-twelve months look like?
- Are the LRTC and six-week calendars posted for all to see?
- What are the big CONUS and/or OCONUS event(s) on the LRTC?
- What is on the six-week and LRTC calendars? Do the scheduled training events build on one another in a logical manner? Are they in-line with the upcoming mission and primarily geared to preparing the team to perform METs needed for the upcoming mission?
- Does the Team have the minimum advanced skill sets required to deploy in accordance with 350-1?
    - Does your upcoming mission require a skill set outside the list in 350-1?
- Are the Team room, storage areas, personal lockers, and latrines clean and/or organized?
- Is every MOS capable of their job? Both tactically and by staff function?
- How does the Team run a range?

- How do they prepare for an upcoming training event? Are there checklists built for training? Or, are they a "last-minute" Team?
- Is there a Team SOP book?
- Physical fitness, does the team PT together, or is it an "individual" responsibility? Is there a specific program (THOR3, Mountain Athlete, CrossFit, etc.) the team is following?
  - Are the ACFT/Unit specific fitness evaluation scores posted?
  - Does everyone look like they could win a non-regulated amateur bodybuilding contest? If so, does that support the upcoming mission?

I cover the assessment period topic in more detail in Chapter 8, Time to Take the Reins (**Handover Checklist** and **The 12 Principles in action).**

1-12.1 Initial physical fitness assessment. Jump in during PT events, but don't take over. Observe how serious the guys take it. Take note who could use more time in the gym, under a ruck, or on the running trail. Lastly, when you PT with the boys you don't have to finish first on the run or the ruck, but you can never be last.

1-12.2 As your first 30 days come to a close, your understanding of the Team's culture, work ethic, strengths and weaknesses will begin to come into focus. If a weakness exists that could jeopardize the upcoming mission, review the LRTC and Training schedules and confirm or deny if that weakness will be addressed. If nothing on the calendar is going to address that issue, look at the time remaining on the LRTC, determine where you can insert additional training, or look at making some adjustments.

1-12.3 More than likely your first 30 days will not provide a crystal-clear picture of what the ODA can and cannot do. Be prepared to conduct a Battle Focused Assessment (BFA) to fill in the information gaps. The BFA is a checklist to help you build a comprehensive assessment to measure your ODA's current capability of the individual and collective tasks for the upcoming mission. Once you have a current snapshot of the Team's capability, you can build out a training plan that does not waste time on items your team does well. This checklist might be Group, Battalion, Company or even Team specific, take some time to dig through the shared drive, engage your peers, and the SGM to find a copy. If you cannot find one, and your peers / SGM have no idea what you are inquiring about, no worries, you can make one on your own.

As a newcomer to the team the mission should be known, and LRTC is most likely built out. If this is your situation, hit the Captain up for the list of METs the ODA needs to be proficient with, and begin to build a BFA. An inexpensive (time / resources) place to start is to quiz the team. Your quizzes should cover the individual and collective tasks that are directly tied into the team's projected mission. Nothing crazy, 10-25 questions in the following areas, or topics relevant to your upcoming mission. You must know where your guys stand before you waste time building an in-depth LRTC.

- MOS-specific tasks (CMF 18 skill level 3 and 4 manuals)
  - Not just tactical tasks, think about all the staff functions each MOS executes. For example, does the 18C know which forms are used when building a pallet, or how to prepare all the flavors of HAZMAT to be shipped?
- Soldier skills (Skill levels 1-4)
- SUT / Ranger tasks, Team SOP book
- Knowledge of MDMP and TLPs
- Core Mission sets
  - Specific individual/collective tasks that support MET(s) needed for your upcoming mission
- Your Team's infill specialty

Now look at the LRTC and 6-week Training calendars for opportunities to insert hands-on/physical assessments, that expose strengths and weaknesses within the required METs. For example, if mounted patrolling is going to be utilized during your projected mission, plan to use most, if not every trip to the range to train and evaluate this task.

- Set up your vehicle task-org board
- Shuffle guys around the inside of the truck every few times out, so they can learn how to operate each position.
- Print off checklists (responsibility cards) by position for pre-mission, and post mission reset /re-cock procedures.
- Incorporate down driver, down gunner, react to IED drills, and don't forget about changing a flat tire rehearsal (does the vehicle have all the required BII?).
    - What does this look like in non-tactical vehicles (if required for your upcoming deployment)?
- Don't forget about the Principles of Patrolling, practice maintaining security during all tasks.

1-12.4 Once you have finished your 30-day assessment, remember that too much change all at once can break a team. Every situation is different, prioritize changes that directly affect mission preparation, and success. Change takes time, and daily discipline demonstrated by you is the key.

1-13 Authorities and Responsibilities. Yes, it is your Team. You are responsible for the daily, hourly, and let's be honest, the minute-to-minute training, coaching, and mentoring[8]. No one can take this responsibility away from you unless you let them.

With that said, at the end of the day, the Captain has Command Authority. Their authority is derived from the President of the United States. US law dictates that officers are responsible for the men, equipment and missions assigned to them. Once you are comfortable with this concept, your professional relationship with the Captain can truly begin.

Something that helped me understand this paradigm was a few concise paragraphs in the *US Army Small Unit Tactics Handbook*, Chapter 5 "Leadership" on page 228. Paul writes,

> NCOs must never lose sight of who is in command and never confuse being in charge with being in command. Similarly, officers must accept and respect the fact that NCOs are in charge of implementing orders and intent.

This issue is also covered in TC 7-22.7. *The Non-Commissioned Officer Guide* in "The Officer and NCO Relationships" chapter. It is a free publication on Army Publishing Directorate and is available in a free app for your phone.

[8] The Training calendar is your tool to manage this responsibility. More on training calendars in Chapter three, pg. 158, paragraph 3-9, Hour by Hour tips, 6-9 Week Training calendar.

1-14 Train, coach, and mentor does not only apply to the NCOs. This falls under CSM Dorsh's Notes on Leadership number 13):

> The relationship between the Officer and NCO is critical. Development of our Officer counterpart (Reverse mentorship) is arguably one of the most important duties as a Team Sergeant.

1-14.1 To me, this means don't throw your Captain in the corner. He only gets eighteen to twenty-four months to learn how to Green Beret. This is the time to develop knowledgeable future Company, Battalion and Group Commanders, so make the most of it. Encourage him to get his hands dirty with the guys, and learn every aspect of the team's daily grind.

We expect him to take those experiences with him as he climbs the corporate ladder. If he gets enough reps with the team, it will be hard for him to forget those experiences when the good idea fairy strikes... one can hope anyway.

You do not know if you will see him again, but there is a chance he might be running the show later in life. If reverse mentorship is done correctly, you can instill a level of trust within the SF NCO ranks that will last his whole career. This trust can supersede the actions of the few bad apples that make it into our organization. He will recognize that the majority of us should not be compared to, and punished with the unwanted few. If you do not put in the effort here, he could take any grudges to his last posting in the Regiment.

1-14.2 Take the time and help your Captain build his "vision" on the ODA. Use your experience, knowledge of doctrine, and experience to influence his products and decisions.

Never Do These Things.

1-15 Never make a promise you're unable to keep. If you are helping one of your guys secure one of the below items, or something similar:

- Lock in a school slot
- Nail down a 6-month college drop
- ETS/retire out of the Army
- Conduct a Group Swap
- Transfer to the B-TEAM, or Battalion to take a knee
- Networking for a Specific SWCS assignment

Prepare them for the possibility that the deal can fall through at any moment. These deals collapse for many reasons that are out of your control.

To avoid dropping the ball on yours, or his end, both of you should sit down and discuss each other's role in this endeavor. Your role, champion his request up and out, get it up to an approving authority. Inform him that you are also tracking 100 other things, for 11 other guys, such as numerous training concepts, and administrative tasks. Tell him that his job is to engage you as often as necessary (make a suggestion like - once a week - or you will be hit up non-stop) to ensure HIS request is delt with in the required timeline.

The advice you give him is the same battle drill you will execute at the company, or Battalion level when chasing down resources for your team. Just as you are tracking 100 things at the Detachment level, they (Company and Battalion) are tracking more.

At the end of the day, regardless of which CSM promised it, be steeled to the idea that any deal made for an individual can crumble.

1-15.1 The above concept is directly tied to creating separation from the boys. You will be tempted to take care of your friends, rather than taking care of individuals for the betterment of the team. Always ask yourself can this action be seen as favoritism. Is this guy getting more than his fair share, or has he performed well enough to truly deserver this action?

1-15.2 Never fight in front of the Team. Clear the Team room, OPCEN, etc., before getting into a heated "conversation" with your Captain and/or Warrant Officer. Once the "conversation" has concluded, and the dust has settled, be sure everyone is on board with the outcome. The Leadership needs to share the same message going forward, regardless of any emotions that were shared during the "discussion". This will help ensure the top three are viewed as a leadership Team with one voice.

1-16 Do these things, *Leadership principle 10: Set the example.* You serve as a leader, mentor, coach, master trainer, and manager. As a leader the guys watch your every move. You must be the standard bearer. If you cut corners, show up late or hungover, fail to adhere to YOUR packing list, etc. they will too. You can never be late, light, or in the wrong uniform.

CSM (Ret.) Flournoy puts it this way:

> Set the example, you truly must walk the line now, and every step forward. When an issue arises with one of your teammates, you must be responsible for that individual. If your professionalism or character is in question you will fail.

1-17 No one is perfect, especially you. You do not know everything there is to know about leadership. Everyone arrives at this job with a different level of preparedness based on their previous experiences, and time spent researching leadership. Even if you were blessed to serve with top notch leaders, do not expect to get it right early on during your time in the seat.

As a Leader, no matter what you do to prevent it, Murphy will get you. You will eventually do or say something that is wrong. You will make a snap judgment, give an incorrect order, make a bad call, or fly off the handle due to stress. Below are a few things to consider when you find yourself on the wrong side of a leadership slip.

1-17.1 **Admitting fault and battling the stigma.** These types of events do not mean you lose all your leadership currency. Either through self-reflection or someone mentioning the event to you, make sure to address it immediately, and take ownership. The guys always know when you are feeding them BS, no matter how hard you try they know when you slip up. If you make any attempt to cover it up with excuses, or act like it never happened your leadership currency drops.

<u>You will not be viewed as weak, or a bad leader if you admit you do not know something, or that you handled a situation incorrectly</u>. <u>They will see strength and personal courage if you admit you were wrong</u>.

1-17.2 **NCOs are never wrong**. This is an antiquated mindset that in my opinion, has no room in our Regiment. Teach your NCOs that being wrong or failing is human, and one of the best learning tools out there. Demonstrate a healthy relationship with failure, and how to best leverage it for self and team improvement.

With that said you will make mistakes as a Team Sergeant. Demonstrate your ability to learn from the mistake and develop an SOP, or whatever is needed to avoid making the same mistake in the future. This job comes with a super steep learning curve and you must be able to quickly adapt and overcome each challenge as it comes your way.

Bottomline, you cannot continue to make the same mistakes. Demonstrate to the team your ability to learn and adapt to your position. Be the leader they need.

1-18 Solutions. Some situations can be rectified by addressing it head on. A simple "my bad" addressed to the team, or to the affected individual can hold some weight. Depending on the situation you might need to identify the root cause of your action(s) and come up with your plan of action/solution before addressing the team.

1-18.1 Sometimes an old school Team bitch session is called for. If you take ownership early, then you can initiate the event. If you try and ignore the issue, there is a good chance the team will put one together. If you have never been the subject of, or been present for one, it is simple. Sit there, be humble, do not get defensive, and take to heart what is being said. In fact, take notes and ask questions.

Ask if the issue is reoccurring and if so, how long have I been doing it. Or am I singling one of you out? How have my actions affected the team? If the guys are willing to speak their mind and tell you how it really is, then they still care and want to bring you back into the fold. If not, you have been counted out, and you can bet a month's paycheck a shadow government has been established.

1-18.2 The guys do not expect you to be perfect. Do what you can, with what you have, and never stop improving on what you have accomplished (good or bad).

1-19 Be situationally aware of your situational awareness. As a Leader and mentor, knowing your operational environment is not just for mission planning and execution. Others outside of your ODA and organization, are watching your every move. You are always being assessed.

1-19.1 We get the opportunity to travel to majestic places in order to represent the red, white, and blue; to serve as ambassadors for our way of life. When presented with these opportunities, be the best our country has to offer. One of my favorite sayings from a boss many moons ago - "Act like you have been there before."

You will have a new guy on most of your trips, and he has not been there before. Hold yourself and the senior guys accountable. Demonstrate the right way to represent the Regiment, the Army, and America.

1-20 SF wisdom. As a mentor, you should have the most time in service, and time on a team. This generally means you have seen and done more than the rest of the guys. Although the above is true, never discount the younger guys' experience from the big Army, other ODAs, and/or civilian life experience.

1-20.1 Your NCOs will count on your knowledge of the ODA's core mission sets, team life, and beyond. If you only know ODA life, i.e., never done SWCS time, staff time, other TDA jobs, you need to reach out. Hit up the SGM, other Team Sergeants in the Company, B-TEAM,

Battalion, and Group Operation Sergeants for their experience. Lean on them to gain a new perspective, the path they followed to those positions contains a treasure trove of valuable information. After you have a better understanding of our community, you will have more information to share during counseling sessions and training events.

1-21 Team Bonds. Personnel that dedicate themselves to prepare for SFAS, pass, and complete the SFQC provide our Regiment with a very unique collection of people. These individuals have demonstrated the required entry level knowledge, skill, grit, and the will to complete a task in the face of adversity. As a Coach, your task is to build a team from this Group of like-minded, and able-bodied individuals.

1-21.1 A method that has stood the test of time incorporates <u>Realism</u>, <u>Basics focused</u>, and <u>Repetition</u> in training. The glue that brings these three essential training elements together is tough training. Hard training brings a team close; teams will always bond through adversity, and that bond, if built correctly, can last a lifetime.

As a leader you will quickly learn who your men are, as will they about you. Everyone's true colors shine through when they are physically stressed (hungry, thirsty, sore, etc.) and mentally exhausted. Although every training concept does not need to be physically demanding, as a

Team Sergeant you need to figure out how and when to incorporate collective training events to build this bond.

When a team is put through a demanding event there is something that happens that I cannot properly define or explain. What I can say for certain is this; one walks away with the gut feeling that no one on the team would ever betray them and that running into a barrage of gunfire to save a mate is no longer a question.

Why? Because they know their brothers would do the same for them. When the individual comes to the above conclusions, the team becomes more important than they could ever be alone. Now, the team can accomplish what appears to be impossible tasks to others, in the most austere of environments, with little to no outside help. The only way this bond becomes stronger is when it is forged in combat.

In my experience the following quote from Simon Sinek is 100% true:

> The highest performing teams on the planet are not the strongest, or the smartest. The highest performing teams on the planet are the ones <u>who care the most</u> (for one another) and are willing to sacrifice for the people next to them.[9]

1-22 Future leaders. From your first collective training event to your last day in the seat, training must

[9]Recorded at reality one-one summit, April 2022.

have the future in mind. You will only be the boss for so long, your seniors will be the next generation to lead ODAs. As Kyle Lamb writes:

> Hopefully, there will come a day when they surpass you. This is the highest form of mentorship.[10]

There are many ways to insert your Senior MOSs into the hot seat. Start small and as they gain experience, give them more responsibility:

- MOS cross training, plan/resource it, run it, hot wash it
- Shoot house reps, quick planning cycle, brief, PCI/PCC, run it, hot wash it
- Convoys to and from training
    - planning, prep, truck side brief, execution, close out activities
- If qualified, plan and run the team's Airborne operation. If done well, volunteer the team to run the next Battalion jump and make him the Primary Jumpmaster
- Develop, resource, and run a portion, or all of your ODA's specialty Infill training / certification / validation event
- When conducting FID training (state side/PMT) pair them with the Partner Force Commander, responsible to help plan and lead a larger element

[10] Leadership in the Shadows, Chapter 29, training leaders, pg. 163.

- Quick option, you are "wounded" during training event X, they have to step up
    - Ensure when you use this method that the team has done a few reps, and your intended replacement has seen it done right
    - Ensure you cover succession of command in the mission brief prior to utilizing this method, the whole team needs to know who is supposed to step up, especially that guy

Place them in any event where they have to make tough calls on the ground. No matter the method you utilize, failure will continue to serve as the best teaching tool when molding new leaders.

1-22.1 Fail early, fail often, and fail forward. Create a safe training environment where your guys want to step up into the leadership position (no, not that safe space bull crap plaguing college campuses). They need to know, at the end of the difficult scenario, no matter how bad they did, this is a teaching moment.

1-22.2 Take the time to execute thorough After-Action Reviews (more on AARs in Chapter 3, **Never be Satisfied**). Although thick skin is required for the brutal honesty an AAR brings, you as the coach need to be ready to provide the tools to rectify the failure. Ask your guy questions to help him self-identify where, and what he messed up. Then provide one or a few methods he can leverage in the future. If time permits put him back into the hot seat and run the scenario again or have him shadow you (or a solid senior) on the next run through. The point being this – always provide a direction for him to follow in order to build on this failure. It is ultimately up to him if he uses it, but you should feel obligated to provide a path.

1-22.3 In my experience, an interesting thing happens when you place your NCOs into the hot seat during training. Either the other team members step up their game, and work harder for their peer, or they do not put out their best work. The latter could be an indicator that you may need to address an underlying issue (such as ego) with the NCO you are training to be a future leader.

As you continue to groom your NCOs as future leaders, and they become more tactically proficient, hopefully they observe the team do everything they can to help him succeed when placed in leadership positions.

This observation, if mentored correctly, will create a new appreciation (for the guys you are grooming) of the team bond I mentioned. He will begin to understand the full weight of leadership, and if treated correctly the team will bend over backwards, even work themselves to death to bring their best for the mission and the leader.

Your goal is to have your NCOs walk away with the following mindset:

> Leadership is not about being in charge. Leadership is about taking care of those in your charge. -Simon Sinek[11]

If they embrace this mindset, they can build the strongest of ODAs when they become Team Sergeants. Lastly, this method will help you and the Captain identify your rising stars for NCOERs later down the road.

1-23 *Leadership Principle 5: Keep your men informed.* As the master trainer for the Detachment, you own the 6–9-week[12] calendar, take the time to build it out. As mentioned repeatedly throughout the book, this doc should hang somewhere in the Team room for all to reference. Hang it near the LRTC so the boys can have some predictability. The calendar will change, keep it up to date and the boys will be grateful. When overseas, hang a copy in the OPCEN.

1-23.1 Do not give the Captain and/or warrant a reason to high jack the hour-by-hour calendar at any time, especially overseas - **NCO creed: Officers of my unit will have maximum time to accomplish their duties; they will not have to accomplish mine**. Sure, kind of a nerd move to include a portion of the NCO Creed knowing my target audience... but the Team could implode if they are forced to completely revamp all of the hard work for a

---

[11] Simon Sinek, Start with Why: How Great Leaders Inspire Everyone to Take Action

[12] As Battalion leadership changes so can this calendar, I was required to maintain either a 6, 8 or 9 week calendar during my tenure as a Team Sergeant.

training event, or a deployment mid-stride all because you failed to do your job, and the Captain takes over.

1-23.2 If you have been around long enough, you have heard the stories of the young/inexperienced, or the weak/burnt out 18Zs losing control. In these cases, the Captains, and/or 180s end up taking over, and the team does its best to right the ship. Do what you know to be right, keep NCO business tight, so the officers can focus up and out.

1-23.3 Here is an example of what happens when inexperience collides with the unknown: We were conducting a JCET, the host nation originally requested training in Advanced Urban Combat techniques, small unit tactics, and basic mountaineering. Once we were in country the assessment period revealed that the partner force was solid with CQB, but not with basic patrolling techniques. They did not know how to approach a target in a rural, or urban areas without the assault bus. We had planned for this and shifted our focus to Patrolling. Our training area, timeline, and equipment supported the required Program of Instruction (POI) shift.

Early on into the patrolling POI, my acting Team leader came up and informed me that we needed to conduct a patrol during limited visibility that evening. I am not sure if he convinced the partner force Commander that this was a good idea, or if he was unable to tell the partner force Commander no? At this point we had only scratched the surface of patrolling techniques, hardly ready to execute a limited visibility patrol.

Our following conversation was not civil. He would not back down because he had already made the commitment for that night. With these factors in mind and wanting to appear as a unified Team in front of the partner force, and more importantly the boys, I got the Team going in the right direction. Begrudgingly we prepared for and executed the limited visibility patrol. Thankfully nothing bad happened, and the rest of the JCET went off without a hitch.

I was a young, promotable E7 at the time. I never witnessed this happen to my Team Sergeants, so I had no frame of reference to go off of. I believe I lost some leadership currency after this event. In an effort to never let this happen again, I reverted to a method that in my opinion (now) is never a good option, dictatorship/micro manager.

Although the guys count on the Team Sergeant to be there to squash or minimize the good idea fairy, this "leadership" approach is never received well. Instead of focusing my efforts on systems to ensure the 180 at the time, or future Captains never highjacked my 6-week calendar again, I executed a full knee jerk reaction, and the team began to suffer. Below I capture some, not all, of what effects took place.

One year after the above trip, the ODA returned to the same country on another JCET. We would be training with the same partner force, working on similar training events from the previous year. The only change was the training location. During the year leading up to the second trip, I began to notice some changes in the team. There were fewer guys who were coming up freely to gain insight on big army issues, SF specific issues, or overall professional development.

Not taking the hint, I began to look for reasons why this was happening. Rather than talking to my 18F, other senior MOSs, or asking the Captain if he had noticed anything, I tried to find faults in the team. My Captain at the time even dropped a hint that a few guys had come to him for time off (appointments, leave, and passes).

Instead of taking this information, doing a quick self-assessment, and adjusting my trajectory, I tried harder to control everything in my reach. Although the details are fuzzy, I am pretty sure I held a team huddle and informed everyone that I wanted them to know I would have the final say on everything they were doing, (micromanaging) starting with leave, pass, and time off for appointments. This attitude bled over into MOS duties, and training events they were planning. I affirmed these items (and many others) were in my lane, never the Captains. To seal the deal, I made some draconian threats to ensure compliance.

Now, we return to the second JCET, the trip's battle rhythm was 5 days of work, and weekends off, not our call, just following our host nation's timeline. Our training area was pretty close to the capital city and being in a permissive environment the Captain and I assessed that leaving a few guys back to watch gear, and prep for the next weeks training was suitable for this deployment. During one of my turns to watch the gear, my senior 18E stayed behind, which I found odd because he enjoyed his "cultural immersion" on TDY.

Not too long into the watch, he opened up to me about the status of the team which turned into a modified team bitch session. He was previously approached by the guys with their grievances in hopes he would confront me with the issues. I believe this was a solid COA because he was one of my top NCOs. He had grown up in the Ranger Regiment, and was a straightforward guy, and about as blunt as they come. If he had something to say, I generally was open to what was on his mind.

He confronted me on all the issues listed above, and many more I am leaving out. It was like jumping into ice water, finally understanding what I had been doing to the ODA. We talked for a long time, I asked questions and took notes.

When the crew returned from sightseeing, I sat them down and we talked about what I had learned. I asked the guys if they had more details, and they did. The rest of the day was spent handling one on one talks with the guys.

Knowing my words could only do so much, I reigned myself in, and tried to let my actions speak for themself. The rest of the trip saw an improvement in the team's demeaner, and we closed it out strong. Upon our return it took time to build the trust back with the men, but we moved forward in a better direction.

1-24 Unfortunate events, ***Leadership Principle #3: Seek responsibility and take responsibility for your actions and decisions.*** Even your rock stars can and will make potentially career ending mistakes. Be prepared to do all that you can for them, you might be the only one on his side. The majority of personnel that make it to Group all start with good intentions. Whatever they did and now are trying to navigate could have started with a good purpose but got out of hand. Remember this when you first begin to help him out. There is no reason to immediately throw him to the wolves without all the facts (Innocent until proven guilty).

1-24.1 Depending on the issue, he will need someone who is not emotionally involved. Someone who can see the whole picture, and a light at the end of the tunnel. He may need someone to keep him organized, keep him on track with all necessary steps, find people to interface with, transportation to and from court dates, etc.

You need to know what programs are available, whether they are Army or civilian, and how to leverage them. You, as the Team Sergeant, are not alone in this, seek the counsel of your peers and Sergeants Major. Someone will have at least one story that is similar and can help you out.

1-24.2 Depending on the situation, this guy could remain on the team. Whether it is during the investigation, court proceedings, or whatever he is going through, let him know that he must continue to work hard, and "Soldier out" of this situation. This old school process is effective and requires that he displays the attitude and actions that demonstrate he WANTS to be here. By Soldiering out he lets the Team know that he has owned his mistake and has learned a lesson. These actions should prove himself worthy to the team, Company, and Battalion. He should already know how this works, but you may need to coach him a bit here:

- Accept responsibility
- Remain humble
- First to work, last to leave
- Executes his duties on time and to standard
- Helps the team wherever he can
- Takes on all the menial tasks (sweep, mop, take trash out, cleans latrines, staff duty, other taskers)

I was this guy a few times in my career, thankfully I had leaders who believed in me, and this method.

1-24.3 With that said, not every guy is worth falling on your sword for. Selection and the SF pipeline can only do so much. They cannot predict the future and know if guys will fall short. Lastly, you need to keep the train going for the team. What if the guy in trouble is your only 18D? What concepts do you have coming up? Do you need to hit up the SGM with an RFF in order to execute training? See Chapter 2 for more information on this topic.

1-25 Time off. As a Leader be sure to have a plan to let your hair down. This applies to CONUS and OCONUS training. The guys will work hard for you, but if you keep them wound too tight, something will break.

1-25.1 When you are OCONUS, and have decided it is the right time to do so, you can have a few brews with the boys.... this is where the adult daycare statement comes in. Obviously, you must maintain control and herd the cats. Work hard, play hard is absolutely a thing, but be careful not to get labeled the "Party Team."

Believe it or not, not everyone in SF wants to party like a rock star. Too much "play hard" will push some of your good guys to other teams. I had a few guys ask to switch teams for "personal growth" or said their interests had changed...The guys who left never said it out loud, but the writing was on the wall.

1-25.2 For CONUS time off concerns, see Chapter 2, paragraph 2-3.2, and chapter 3, paragraphs 3-4.1, 3-4.5.

1-26 Accountability. Your guys will get after it during their off time. I strongly suggest implementing a

GOTWA[13] SOP for the boys. A simple 5Ws over text can help keep track of those who head to the mountains, the lake, etc. on the weekends.

1-26.1 Example: One of my guys took off to the Sangre de Cristo Range during a long weekend. He left a GOTWA with a team member and took equipment for a possible overnight on the mountain. When he got stuck on the climbing route due to bad weather, he was unable to notify anyone because his cell phone did not have service. With the GOTWA in place, I had the basic info to call Search and Rescue on Monday, when he did not show up for work. At the end of a very long day, all was well because he went prepared (food, water, clothing system, bivy bag, etc.) and I knew where he was.

1-27 Conducting an audible. If you brief the boss, you will be wearing a military uniform when conducting training, stick to the plan. If the topic of straying from the briefed plan comes up, conduct IPB prior to committing to that change. Fully understand your operational environment.

1-27.1 Example: We were conducting detachment level Cold Weather Training at Loveland Ski Resort. The ski passes we were issued from the Group S3 came with specific instructions: "All personnel using the passes MUST be in uniform." We were scheduled for two weeks

[13] Where is the leader **Going, Others** he is taking with him, **Time** the leader plans to be gone, **What** to do if he does not return on time, **Actions** by the element in the event contact is made while the leader is gone.

and were not expecting any Command visits. The boys hit me up about dropping military clothing and rocking civilian stuff. I decided against it, and the decision was not liked. No shit, within a day or so of that decision, the Group Commander just happened to run into us on the side of the mountain. This "chance" encounter ended on a positive note because we were doing the right thing.

He must have communicated his impression of the ODA down to the Battalion Commander, because the remaining time he was in charge a large majority of our mountain-based training was approved.

1-28 ***Leadership Principle 7: Employ your team in accordance with its capabilities.*** You are the principal advisor to the Detachment, AOB and SOTF Commanders for the ground truth regarding your team's capabilities. Do not put your team in a bad situation that you are not prepared for. As you know, or will get to know, your detachment executes missions that have effects on the strategic map. If you know without a doubt that your team is not ready to conduct a mission, then speak up.

1-28.1 If your words are not heard, and the mission moves forward, your job now is to prepare the team the best you can, and get the job done.

1-29 Acceptance. You and the team will dedicate many hours to develop training concepts. In order to get out the door and conduct that training you MUST accept the following as fact: Admin is king. Your days of implementing the Five D's of Dodge ball[14] (Dodge, Duck, Dip, Dive, and Dodge) regarding administrative tasks are over.

1-29.1 Your Team will get more training opportunities if you make sure all the admin boxes on the trackers are green. A solid battle rhythm, discussed in the next chapter, will help you on this front. <u>You are 100% responsible for all things administrative on the Team.</u>

1-29.2 Every unit has their own culture and focus points regarding administrative items. On top of that, every command team (Company and up) will have their focus points on top of those. Get to know what those items are. Armed with this knowledge, you now know what you have to do in order to thrive as a team. Do not let your ego get in the way. If your only interactions with the bosses are hostile, then your ODA will slip on down the OML for everything. I am not suggesting bending over a barrel, but choose your battles carefully.

1-29.3 Example, if the SGM is switched on in the admin world, make sure he has zero reason to engage you in this area. If you are early and submit your best work, then he will feel more obligated to fight for your team's needs.

If you happen to have a SGM who is switched on in the admin world, lean on him. This is the guy who can help you step above your peers. Think about it, if we are all

[14] Dodgeball: a true underdog story (2004), movie quote, Patches O'Houlihan.

physically fit and tactically sound, what else is there to measure you on when being looked at for a promotion to SGM?

1-30 To close out this chapter, all the above topics should be considered when preparing yourself to serve as a Team Sergeant. But... You must remain true to yourself. Apply the above general rules of the road with your personality. Restrain yourself if you are an extrovert and super sociable. Push yourself to be available for the guys if you are an introvert. Anyone can fake the funk for a period of time. But when life is difficult (in training or in combat) your true colors will always shine through.

No matter what you need to reign in, or push through, keep the above in mind, and your team time should end with success.

2

On Admin and Counseling

2-1 Admin is King. Stay on top of this, it keeps the SGM happy. When the SGM is happy you have freedom of maneuver.

A balance must be mentioned here. Yes, you own the admin on the team, but do not chain yourself to that desk. As I mentioned earlier, people in our ranks have issues following a guy that won't do the work with them.

You need to be at the training events because you are the master trainer. If your guys are running specific POIs, you must monitor what the team is doing. Are they implementing the team's SOPs?[1] Is the risk assessment being adhered to? There are many reasons to be present at training, make sure it is a top priority for you.

> An organization does well only those things the boss checks - GEN. Bruce Clarke[2]

2-2. Battle Rhythm. Having a battle rhythm established will help you maintain accountability of the team's administrative requirements. No matter where you

---

[1] SOP book discussed in Chapter 6

[2] Guidelines for the Leader and the Commander, GEN. Bruce Clarke, Chapter 5, pg. 30.

are within your training cycle, these ankle biters will always pop up. Your battle rhythm must be simple to use. If it is too complex, or rigid, you will not be able to keep up with it during fast-paced training periods.

2-2.1 In my opinion, the first step in figuring out your ODA's battle rhythm is knowing the B-TEAM's (which is based on the Battalion's) schedule. Training meetings should be a familiar concept to you. You have either sat in for your Team Sergeant before or know this is the meeting where all of the bad news (taskers) come from. More crucially, it is the platform for the detachment leadership to voice the needs of the ODA.

If your Group is doing it right, they will follow CSM Dorsh's model:

> Notes on Leadership 9) **Incorporate a strict and disciplined work week calendar:** Monday– prep/admin, Tuesday through Thursday–train day and night, Friday – recovery/admin. It's simple to understand and implement and proven to work. Take a look at when you are holding Company/Battalion training meetings/Command and Staff – is it Tuesday – Thursday? What message does that send to the force and to our leaders at echelon?

2-3 Admin weeks, better known as **white space**. These weeks consist of four or five office days in-between training events. If you just finished six weeks of SFAUC training, you and the guys need at least one week, possibly two before you jump back on the training wagon. Emails alone will take a week to sort through. Not to mention all the other stuff the SGM is harassing you for.

2-3.1 White space is the time to apply your admin battle rhythm. Below is a sample list of topics you and the team will focus on:

- OCONUS mission products tracker items/JCET timeline work
- S1 items (NCOERs, awards, pay issues, language tracker)
- Training concept tracker items
- Jump logs, ensure all are current jumpers, JMs are current, begin/continue to coordinate the next jump
- Weapons maintenance
- DX un-serviceable items
- Complete the one-off mandatory annual online training certs that fall outside your established block
- Medical readiness/doc appointments/sick call
- Not a bad time to try and schedule 18D credentialing
    - Double edged sword here, if your 18D is taking off for credentialing, ensure he has knocked out any of his admin requirements prior to departing.
- Property book maintenance
- Ammunition draws, or dunnage turn in
- Preparation for the next training event
- Creating/updating the Team SOP book
- Team PT/fitness assessments

2-3.2 Don't forget to use these short weeks to plug in those missed Holiday's. Lastly, these should be short workdays, get the boys home to their families.

2-3.3 Below I have provided an <u>example</u> battle rhythm during an admin week/white space:

- Monday, open the week with a team meeting. Lay out the week's priorities, and which day to focus on those priorities. Use Monday to review the OCONUS mission tracker. Prioritize products that have a short suspense. Then tackle the medium to long range items.
- Tuesday could be your S1 day. NCOERs, awards, pay issues, Jump logs, language tracker-know who is due/schedule training or testing.
- Wednesday is training concept day. Which concept(s) still need products completed at the ODA level? Where are the other concepts in the approval process? Time to track down resources like: Affordable lodging at the training location, money for equipment or training, support MOSs, vehicles, etc.
- Thursday can be for the boys. Give them time to hit the doc up for that nagging injury. Time to get a physical started or completed. Work on a POI for team cross training, rehearsals, etc.
- Friday morning, close out meeting, review tasks that were assigned. Prioritize and execute incomplete tasks that directly affect the following week.

- If any of the assigned tasks - regardless of the day of the week - do not take all day, pivot back to medium and long-range product production (concept development, OCONUS mission prep).

2-4 Connectivity. The admin work does not end while you are training. Prioritize and execute all the admin you can during "white space" admin weeks. This gets you out to training with the boys or ensures you can enjoy leave with your family.

Speaking of leave, accept the fact that as a Team Sergeant, you are always on the clock. When the team takes leave, you are never fully on leave. Your phone, and laptop are always nearby. Also, no matter how good you are at the admin tasks, you are never ahead. Ensure you can handle some, if not all of the below tasks at home, or on the road.

2-4.1 Unless you are deep in the woods, under water, falling from the sky, or high up in the mountains executing pure awesomeness, it is rare to be away from a computer. What I am trying to get at is you have the time and technology to remain on top of things while executing most training events.

Not too long ago, when we would return from a weeklong trip here, or two weeks there, I would dread opening the email machine. Now, if you have a Wi-Fi puck, or a smart phone, you have connectivity. This capability is a blessing and a curse for obvious reasons.

The ability to chip away at NCOERs, awards, future training concepts, and emails a little each day warms the soul. This will help you stay ahead of the power curve before you reach your next admin week. On the other side of the coin, you are easily reachable for all those people and things that distract you from training.

2-5 Tracking methods. There are many requirements to keep track of and they all have different hit times. Daily, weekly, bi-weekly, monthly, semi-annually, some events and tasks overlap, most are spread out over months, or show up once a year. Except for a few items, these requirements will never be aligned in the same timeframe. Example, everyone's OPI will be spread across the calendar. There are many methods available to assist you in this endeavor. Remember, your tracking system is only good if you are consistent in using it. Here are a few options:

- Green notebook
- Phone calendar reminders
- Outlook calendar
- White board(s)
- OneNote
- Old school day planner
- Mega excel tracker
- A combination of the above

2-5.1 **The Green Notebook**. Hand jammed notes, SO. MANY. NOTES. Company training meetings, random conversations at Battalion or other locations, phone calls, texts, emails (these are the worst). I used the following method with my green notebook. When I was consistent, not much was missed.

At the end of the week, review your notes. Cross out completed activities (complete a task, cross it out, and receive a small hit of dopamine), consolidate the pending tasks, and move those forward to a new page. This does a few things, one, you move to a fresh page which cleans up the list. Two, you automatically re-prioritize the remaining tasks. Three, by re-writing the list, you commit to memory what needs to be done, just in case you are without your notebook.

Come Monday, before the team meeting, review your admin and concept tracking system(s), review the notes from last week... <u>Prioritize</u>, <u>Brief</u>, <u>Execute</u>, and <u>Monitor</u>.

2-6 Team meeting agenda, have it set ahead of time because the guys will lose interest if you are all over the map. These meetings should be kept as short and concise as possible. Don't allow the guys to dive headfirst into rabbit holes. Sure, that topic does require a longer conversation, but this is not the venue. This goes for you as well, if you get long winded on one or more topics during a meeting, you will lose the guys quick.[3]

---

[3] Note: If you have a lot of detailed instructions for a task, this might fall under NCOPD. Do it right, set the team room up, or reserve a classroom. Prepare read ahead material and give a block of instruction.

2-6.1 If the team is TDY for training, let them focus on the event. Dedicate the evenings to hold a meeting if required or go point to point with guys that need to knock out a requirement.

2-6.2 The OCONUS battle rhythm will iron it's self out within a few weeks depending on the mission. Don't be surprised, you may start with two meetings per day just to keep all parties on track and check the progress of critical tasks.

2-6.3 If your team is split, or fully decentralized, do what you can with who you have available to make things happen. Have a sure-fire method to get critical information to the guys. A team chat is nice, but they can become a dumping ground for wildly inappropriate messages, memes, and gifs. I am not opposed to that humor, but it could cause some guys to ignore the important info being put out.

2-7 Admin Tracking. Now that you have a starting point on HOW to keep yourself organized, what needs to be tracked? The below list is a good place to start, but I am sure things have changed since my time in the seat and will change again while you are in charge.

- Annual online training
    - I gave this task to my 18F. He built the team's book that held all the physical certificates, maintained a folder on the I drive, and a corresponding excel tracker.

  - Do not waste time trying to figure out what the requirements are, have the B-TEAM request the official list(s) from higher.

- Range Safety Officer (RSO) cards
- Ammunition handler
- Military license (who is HAZMAT qualified?)
- Unit Movement Officer (UMO) certification
- Passport expiration dates (civilian/government) and where do you store the passports?
- GOVCC expiration dates
- DTS certifications, and voucher submission /completion status (post team trip, or individual trips)
- 18D medic credentials
- Clothing sizes (boots, hats, gloves, cold weather gear, etc.) you never know when Uncle Sam will give out free "military grade" clothes
- Service member (SM) arrival date to the team, Battalion, and Group
- Projected PCS, ETS, Retirement dates
- NCOER thru dates
- Counseling schedule
- Physicals, know when each SM's expires and make time for guys to redo them. If I could do it again, my 18D would track this.
- Awards (PCS, ETS, retirement, impact, etc.) status, another 18D responsibility

    - 18D not only tracks team award status, but is prepared to deliver a professional development class on how to properly fill in a DA Form 638
- Skills Tracker, completed NCOES, Advanced skills, big Army schools, etc., Not just a banner to showcase your team's awesomeness. This document will identify capability gaps IAW 350-1, helps you build your schools OML, and aids in identifying individuals who require schools IAW DA-PAM 600-25 and the SF Professional Development Model
- SRP packets, see chapter 3, paragraph 3-4.6
- DA 330s with language scores, type of test, test date
- Jump status/Jumpmaster currency/potential pay hurts/pay losses

2-7.1 Throughout the book I am honored to include MSG Enrique Longoria's input. I met this fine American when we were both Chief Instructors at SUT. His contributions within the book are identified with an (EL) at the start of his paragraphs. In the next few pages, he puts all of the above information together:

> (EL) Along with Annual Training trackers, I was a big fan of visual trackers posted on my desk for all to see. I kept trackers for:
>
> -Alert Roster
>
> -Alpha Roster
>
> -Advanced Skills

-Language

-Counseling and NCOER dates

-Jumps

Not only does this keep you on track, but it allows the individuals to help you manage as well. There are so many short notice emails or requirements that you will need to handle. If you were out, for whatever reason, your Assistant Operations Sergeant has something he can reference for continuity, as opposed to searching through the Team I: Drive trying to find it.

Team meetings are essential. When I could, we held a Team meeting two times a week. Monday-open, and Friday-close out. In these meetings MOS's were responsible for digging into the Command and Staff slides, with the implied task of briefing the relevant information that affects the ODA.

This exercises the staff functions of the MOS's, and builds trust with your Team Leader by the men displaying that the ODA is tracking all aspects of their job. Lastly, this serves as development for the men. They learn what the command is looking for, tracking, and what is expected of them as an Operations Sergeant to monitor and manage.

With that said, you, as a Team Sergeant, need to supervise this and ensure that the MOSs are capturing everything that needs to be addressed. This may seem repetitive, but a leader must spot check.

2-7.2 I provided an additional post Team Sergeant's view and methods for a reason. Although our team time was separated by a few years, and we grew up in different Groups, the general idea is the same.

2-8 Below I am going to open up about a situation I found myself in that I fully own. My hope in providing this story is to drive home the point that ignoring the admin requirements can jam you up in a bad way.

2-8.1 At the time, I was the NCOIC of the Mountaineering course, this was my first post Team Sergeant time assignment. The Cadre team and I were gone a lot for training or running courses. With this fact, I had let my attitude revert back to that of a senior team guy and did not focus on my responsibilities as the Team Sergeant. I had not balanced my time between training with the men and completing my administrative duties correctly.

This led to a robust list of deficiencies that the SGM and Commander were no longer going to ignore. The big-ticket items were:

- Jumpers who were not even close to current, some had gone 18 plus months without exiting an aircraft in flight
- Multiple late NCOERs
- Late PCS and retirement awards
- Late training concept products

In this scenario, the chain of command did not care that the Army just paid $10,000+ on a training event we were currently conducting. Lucky for us, the course cadre were empathetic, and conducted some late-night study halls for my guys that missed 24 hours of training.

So, there we were in Ouray, CO a mere six-hour drive away from Fort Carson in the middle of a bitter cold winter. The training was five days long, Monday through Friday. I received a phone call from the SGM early on in the week. He informed me that I would send my non-current jumpers back to strap-hang on a jump. The pressure to get guys under canopy was based on Group's jump logs had been audited, and the men were going to lose pay.

On top of the jumping issue, I was instructed to have all my delinquent administrative products in by close of business Thursday. The afternoon I received these instructions I prioritized my jumpers and set up a plan for their travel back to Fort Carson. Next, I sat down and reviewed my administrative workload. I was embarrassed of my ability to procrastinate, and what I had let slide. I remained in my hotel room the rest of the training week.

2-8.2 After this failure within my administrative duties, and missing out on critical training, I re-prioritized my battle rhythm. Along with the paperwork side of the admin world, I became slightly obsessed with the different Jump rules. I needed a solution to this problem.

Background on why many of the cadre were so behind on jump status: A handful of my guys arrived to the school house as pay hurts. I did not do my job and scrub each man's jump log when they signed in. Next, as I mentioned above, we were gone a lot. TDY within the state of Colorado, and a few trips to Wyoming, an average of 29 weeks per year. Some of the cadre were pushing close to 40 weeks per year.

Additionally, just like anywhere else in the Regiment, my guys were attending advanced schools, NCOES, signed up for college, and taking leave with their families in between courses. To make things worse, the weather in Colorado constantly created no jump conditions. Our courses always overlapped with the "good weather" months, so we were never around to get manifested.

The solution to our jump issue I came up with was to get everyone's jump status current during our off season (late September through December) and put everyone on Jump Rule Three. This required a detailed plan, briefed to the AOB leadership, and the Battalion Commander, who would sign off on each SM's Jump Rule 3 memorandum. The cadre were fully on board with this plan, anything to avoid a repeat of the Ouray trip.

USASOC regulation 350-2 has since changed in a good way to help us out on this front, but the lesson learned is you have to be creative to fix your own problems. Be ready to do the work, know the regulations inside and out, or know where to find the information to solve your problems. Nobody is going to care about your guys more than you.

2-9 Trust up and down the chain of command is important. There will be times when you have to inform the SGM that the team is not in a position to execute any additional taskers that come down the pipe. If you have built a solid foundation of punctuality in the admin world, he should fully understand when you make this request. After my above self-induced situation, it took a few months of consistent work with my admin responsibilities, but I Soldiered my way back into his good graces.

2-9.1 To help navigate the above situation, lean forward in the foxhole, and produce a detailed troops to task. This product will show the SGM, in detail how each man is currently, and is projected to be occupied with work, schools, taskers, or deployments. It is a fantastic visual tool to prove you truly do not have a guy for that random tasker that just came down.

2-10 Counseling, *Leadership principle 4: Know your men and look out for their well-being.* There is no way I can cover every counseling scenario in this book. Below I cover the big-ticket items you need to prepare for, and some situations I had to handle.

2-10.1 First thought on counseling, it is an art. Your preparation and delivery will be based on many factors. To list a few:

- Your temperament (more on temperament in Ch. 5)
- Past experiences (as counselee, or counselor)
- Types of counseling – initial, quarterly, negative, positive, NCOER review and signing, formal/informal, etc.
- Location (team room, car ride, bar, wood line)
- Relationship to the SM – long-time friend, team transfer (seasoned guy), instructor returning from SWCS, or a new guy from the SFQC, etc.

2-10.2 Next, most counseling sessions are a combination of using an attention getter, then an application of teach, coach, and mentor. The attention getter is not a literal kick in the ass, but a figurative one... or is it? FM 22-102, Wall to wall counseling, check it out sometime. Once your guy knows the topic of this counseling session, the critical part comes into play. <u>You must follow through with the appropriate amount of Teach, Coach, and Mentor.</u>

For the above, and following information on counseling to work, you need to adhere to this maxim:

> "Curious before Furious." – LTC (Ret.) Curtis Price[4].

What does that mean? Expend some effort to figure out the root cause of the behavior. If one of your dudes just recently started slacking off at work (arriving late, not shaving, poor PT performance, missed deadlines, etc.) find out if everything is ok at home. Once you know what it is, do what you can with what you have to support him fixing the issue.

2-10.3 Here are some possible examples of the above method within specific counseling sessions:

- <u>Initial counseling</u> with the new kid straight from the Q course. Attention getter – You need to clearly explain that our team of 12 has an impact on the world stage. If explained correctly, this should open the SM's eyes to the importance of what we do, and everything he does, or fails to do, can affect the team's overall mission.

  Then execute the arm around the shoulder, teach, coach and mentor. Explain how small the number 12 really is. Tell him that we need him to begin mastering his MOS, and cross train in others. Describe how we all lean on each other to complete

[4] Coworker of SGM (RET) Bill Hanes, Assigned to 2nd SWTG(A) circa 2021-2022.

our mission. More on initial counseling in paragraph 2-11.

- Negative counseling events. The following is geared to capturing an event on paper after an incident. There is never time to stop training, whip out a laptop, and begin hammering away at paperwork. Yes, you immediately deal with a safety violation, a major issue, behavior, etc. verbally, then you attack the paperwork (if required) at the end of the training day, or event.

  After the training event, AND you have calmed down, ask yourself is the event paperwork worthy? Is this the SM's first ever (big) offense? What were the results of his actions? If it is worth the time, capture the pertinent facts on a DA Form 4856. The facts of the event become your attention getter. Your delivery of the facts should cast the brightest light onto the severity of his actions. Also, it will set the stage for a productive counseling session. There should be zero question to why you two are having a heart-to-heart conversation.

  After he understands why you are sitting there, implement Curious before furious. Dig into the how, what, and why he did what he did. Be prepared, you may find out that he did not have all of the information or training he needed to safely, or fully complete the task. If this was the issue, are you at fault? Dissect this information, you may have to inform the SM of your mishap. Use this moment to demonstrate how to own a mistake, and to develop a plan that ensures this does not happen again in the future.

More than likely though, you will discover he does not have the life experience, maturity, or the capability required to operate in the environment/situation he was placed in. Knowing this will aid in the development of the SM's plan of action. The course of action block is where you capture the teach, coach, and mentor plan. This plan should align with the issue at hand. Squash your desire to implement a knee jerk reaction that makes you feel better but does nothing to promote development.

Without teach, coach and mentor, the SM learns nothing, and possibly becomes a liability to the ODA. Sure, being a GB requires the ability to figure things out on our own. In fact, giving our guys assignments, orders, and tasks without detailed guidance is a big part of professional development. But, depending on the issue at hand, you need to build and implement a plan to square your guy away.

2-11 Initial counseling, the importance of, and what I believe should be included. Before I dig into those two important topics here is my first tip - review the following, and any other documents you deem worthy, to get an initial snapshot before you sit down with your counselee:

- Last three to five NCOERs (DA 1059s if an 18X)
- SRB
- DA 330
- Jump log

Use this information to fill out each man's 5-year NCOPD model (see figure 2:1).

Your counselee's SRB will most likely be lacking. Be sure to coach him on the importance of keeping his records up to date, and have him produce his graduation certificates / deployment orders / awards / college transcripts / etc. After the counseling session get him up to S1 for updates. For more information regarding a detailed SRB scrub see paragraph 2-21 and Chapter 3, paragraph 3-4.6.

2-11.1, The importance of or better said the WHY behind initial counseling:

1. List out your expectations for/but not limited to - MOS, staff functions, specialty skills, team SOPs, understanding our core missions, timeliness, physical fitness, etc. <u>Without this information</u> the individual, and ultimately the team have to play the guessing game based on how you react when you find something not to your liking. When the team has this road map laid out, they can truly focus on learning and mastering their craft.

With these standards written out you can refer back to this document during quarterly counseling sessions, or if a negative situation arises. Lastly, it is difficult to throw a dude's ruck sack in the hallway without an initial counseling (and many more) in his packet.

2. The initial counseling session is one of the best ways to start building trust. Whether you are the new boss coming in, or an old hand bringing on a new guy, an initial counseling session is a good way to begin building a foundation of trust. If you are personable, open and honest with them, they should reciprocate. Once the men believe you have their well-being in mind, you will begin to see signs that trust is being built.

2-11.2 How to start the conversation and which topics I believe should be included. The following template is not the only way to execute, but it worked well for me.

A solid way to start is to tell them a little bit about yourself. Not a full historical rundown, just enough to cue him in on what to cover: Married with kids, time in service, time in group, previous MOS(s), previous units, and maybe a hobby. Now that he knows a little about you, and he understands that this isn't an interrogation, he should open up.

Other topics to cover early on in the conversation:

- For new guys what is their motivation for being in the Army, and in SF? For seasoned guys, what is it that keeps them coming back? If you know what drives them, you can look for opportunities to fuel their driving force.
- What are their current concerns (career, immediate and extended family, buying a house, etc.).

2-11.3 Next up, layout in detail your expectations for:

- ARSOF attributes, team culture and rules of the road
- Duties and responsibilities by MOS / staff function
- Physical fitness standards
- Timeliness (product production, hit times, information sharing – as in bad news does not get better with time, etc.)
- Language and AOR cultural knowledge
- Acquiring and maintaining Specialty skills
- Understanding our Core missions
- Maintaining individual and team equipment

If your team SOP book is complete, you can reference specific sections on the counseling form in an effort to reduce extra typing.

Additionally, this is where you can define your expectations for Most Qualified, Highly Qualified, and Met Standard NCOERs. Ensure he knows who his competition is.

If you are planning to capture your team's cultural expectations on the initial counseling form, you have options. The DA form 4856 is limited on space, for example my first initial counseling was a generic DA 4856 that simply stated, "see attached word document." The word document was full-bodied to say the least.

The other option is to be short and to the point. Focused on the ARSOF attributes, rules of the road, and ODA basics like "don't air out the team's dirty laundry/speak ill about the team to anyone outside these walls..." I have used both, but to be honest the robust option should be the go-to for new guys coming from the SFQC.

2-11.4 Following your expectations, shift the counseling session's focus to his SRB and the Five-year NCOPD model (paragraph 2-16 in this chapter). This topic is another way to demonstrate you have their wellbeing in mind (see paragraph 2-21 for a detailed list regarding SRB scrubs).

Example, if you receive a seasoned guy (returning to Group from SWCS) and you see that he is not JM qualified, or that he has not completed DLC for SLC, you can incorporate that into the preferred school's conversation. Tie this information to DA PAM 600-25 and show him where it resides within the regulation.

Most guys have a career path they want to follow. This is your time to help him stack schools by priority to meet those career goals. Inform him that JM, or DLC comes first then we'll talk about advanced schools. The SF

Professional Development Model (PDM) is a good visual reference for this part of the conversation.

While talking about his school goals, if he has not already laid it out, ask him about short- and long-term goals. I broke these into 1-year, and 3-year chunks. His 1st year goals should align with what the Team is preparing to execute, and the 3-year goals should be aimed at his career path goals.

Additional topics to cover:

- DA photo review, once every 5 years, or after specific new award(s), and promotions (yes, still important even if not used during the evaluation boards).
- Jump log, check to see if his last unit closed him out. Confirm when his last jump occurred, does he need a BAR/JM refresher, and add him to your tracker.

2-11.5 MSG Longoria's input:

> Following my initial counseling, which was 30 days after I took the Team, I told them that in 60 days we'd be conducting Team physical gates. As such, they'd need to be prepared to meet the ODA standards (as defined in the initial counseling). As an example of our Team's physical preparedness, we tested on the following:
>
> -APFT with 7x dead-hang pull ups (ACFT now)

-UBRR

-5 Mile Run

-12 Mile Ruck

-1km / 2km / 3km surface swims (for a dive Team)

-5th SFG / Air Assault Obstacle Course (for time)- this was more for fun, but you can identify shortfalls in your ODA's physical capabilities.

I posted the scores of the above events on a Tracker in the Team room. I did this for two reasons, first, to instill a sense of internal competition, and second, to sell the Team to the Command or anyone who enters the Team room.

It goes without saying that it served to encourage esprit de corps and reinforced a sense of dogged athleticism. Life's a competition, I applied this same martial philosophy to other areas. For instance, shooting statistics for the various critical task evaluations (CTE) we conducted during ranges were also tracked on a board for all to see.

Although Enrique's input primarily focuses on what his physical fitness expectations were going to be, it is a solid example of the detail you should provide during your initial counseling sessions. He also included an example of his Team's culture by being overt with posting the scores for all to see.

Bottom line, if you want your guys to succeed, you must communicate clearly ALL of your expectations.

2-11.6 For inspiration regarding initial counseling, see *Leadership in the Shadows* by Kyle Lamb. His Counseling chapter is chalked full of helpful tips, to include a full ODA initial counseling.

Additionally, grab a copy of *Always Endeavor, a developmental guide for in extremis leaders* by Colin Greata. Look in Part III, page 177, The community's expectations, your initial counseling statement from the SOF community, and Mindsets for approaching teamwork (page 187).

2-12 FNG. Whatever your initial counseling ends up being, an introduction of the new guy to the team is necessary. Why, because within a team construct, one feels like an outsider until they are brought into the fold by all members on the team.

2-12.1 Many Teams have long-standing new guy probationary periods, and traditions which address the above in their own way. For the purpose of this book, I am only asking you to review these traditions and ensure at some point the new kid is brought onboard. Of course, he must prove his worth. I am not trying to sell some new age hippy stuff. The point is this, extending trust upfront, and taking time to show him the ropes, will make him feel like he is value added. This instills the feeling of belonging and being a part of something more important than himself. This leads to better performance.

2-12.2 Lamb recalls his first day on the job, and how his boss received him in his book *Leadership in the Shadows:*

> Jon had accepted me onto his team with open arms. He immediately made me part of the team and pushed me to take responsibility. He told me I needed to "do what needs to be done." Get out there and figure out what can be done to be better, whether it be through training, tactics, or technology. If the trash needs to be dumped, get on it. Your rank and position do not make you immune from work. No task is unimportant.[5]

2-12.3 Holding a team huddle with a quick "hey, this is Rambo, the new guy" should be part of the process, but it is not enough. Take the time to do a face to face with each MOS. When you are with that MOS, briefly cover what this individual does for the team. Then give your instructions to that MOS to ensure the new guy feels included. If I could re-introduce all of the new guys that came to my team, it would have gone something like this:

- "This is Andrew, the senior 18C, he will get your team gear handled." Then take the next step, "hey, Andrew, by the end of the week make sure he has his complete initial issue, CIF, and SPEAR. Let me

---

[5] Kyle Lamb, Leadership in the Shadows, Chapter II, page 47.

know if you have issues at the SPEAR warehouse, he needs his equipment for next week's training..."

- "This is Dunny, the Senior 18B and he will help you set up your weapons. Dunny, make sure John J. sets up his primary weapon with the appropriate optics for our upcoming range. Then work with him on kit set up, to include his pistol belt. Show him the team SOP regarding minimum kit loadout..."
- "Yancy is our Senior 18D, he will double tap your in-processing checklist, and get your IFAK set up in accordance with our Team SOP." Follow up with - "Yancy, I want you to ensure his SGLV, SMIF, and burial worksheet are square before next week's range..."
- "Eric is the 18F, he will work with you on the country study for our upcoming trip. He is also my 2IC, if you need time off, for a doc appointment, house closing, or other type of issue, and cannot find me, hit him up." "Eric, get a copy of all his annual online certs for our book. If any are out of date or missing, set him up with the links and ensure he gets them done before next week's training..."
- "The other Eric is our Senior 18E, he will help you gain access, and navigate the shared drives, portals, etc." "Eric, get John J. up to speed with his radio since we will be using them next week..."

- "This is Chief Huestis our 180. He is the Chief of staff for the Detachment. He can help you fine tune your staff role on the team as time goes by. Additionally, he can help you dive into the big picture items our team is involved in. Lastly, if you have questions about the long-range calendar he owns this product, hit him up."
- "This is Captain Brown your Senior Rater, he and I will sit you down very soon for initial counseling."

2-12.4 You should review the task list you just gave him, as we all know the first few days and weeks on the team can be intense - set him up for success. Show him all the important documents hanging in the team room (6–9-week calendar, LRTC, MOS task boards, Vehicle task org board, etc.). If your Team SOP book is done, this is a good opportunity for him to start reading it.

2-12.5 Of course, one of the above MOSs will be his Senior. This introduction is pretty important as he will be a major influence on the FNG. You should prepare your Senior(s) prior to receiving a new guy. Take the time to layout the topics in Chapter 5, and anything else you believe to be important within the leadership realm.

Now that I am thinking about it, this should be part of your NCOPD plan for all your Seniors. Always be preparing for the future, make sure your guys have the tools before they move up into the hot seat.

2-13 What about you? The Captain is your rater; he owes you an initial counseling. When you arrive, the current Captain should have this prepared.[6] If you are receiving a new Captain out of the SFQC, be prepared.

Some Captains are squared away, and they will approach you about this subject. Either someone sat them down and covered this topic, or their past experiences set them up for success. Or, they are not prepared, and do not know how to approach you on this matter.

2-13.1 If you are facing the latter situation, you will need to prep the environment for a good dialogue. Nothing crazy here, inform him that he owes you an initial counseling. When he is ready, select a time and location where you won't be bothered by the team, or the B-TEAM. Don't let him ask you for your last three NCOERs and SRB. Have those ready for him to review, as he builds your initial counseling form.

Hopefully he is resourceful, and sits down with his rater, and the Company SGM to develop your initial counseling. At a minimum he should focus on the 18Z's Duties and Responsibilities for Planning from MDMP resources. He should understand your job as down and in, while his is up and out. He should explain his understanding of how both of you do these different tasks, while never losing focus of the Men or the Mission. He should want to get to

---

[6] See Appendix IV, Pg. 289 for my initial counseling from my detachment Commander.

know you, just like you do with the boys – Are you married, kids, how long you been in, short-and long-term goals, etc.

If he does not follow the above, or just seems lost in the sauce, help him out. Reverse mentorship, as mentioned a few times in this book, is part of your duty description.

With either option, use this time to ask him questions. Start small, see what he knows about your job. Ask questions about his job, see if he fully grasps his role on the ODA. This is the time to fill him in regarding the team's battle rhythm/SOPs, and his responsibilities within those. This is a good time to discuss, and clearly label NCO vs. Officer "business".

2-14 Quarterly counseling is your tool to help train, teach, coach and mentor your guys to be their best self. Although you should be deliberate when conducting this type of counseling, it does not have to be formal, nor done on a DA 4856. If you are prepared (agenda in hand) you can do this where and whenever you have a few minutes (the range, traveling to training, the bar). Be sure to have a method of capturing the dialog, then start a conversation.

2-14.1 When conducting quarterly counseling, ask questions and listen to your guys. By asking leading questions the counselee becomes more involved with his future. He should create eighty percent or better of the plan of action. If you spend the whole session dictating how to execute tasks, then you are micromanaging. Also, counseling helps capture NCOER bullets. It is way easier to remember what happened 1-3 months ago, vs. 6-12 months back.

2-14.2 When you are conducting quarterly counseling, or any counseling for that matter, be sure to layout a clear intent. Not every session has to cover all the following topics, but once a year all of the below list should be covered:

- How they currently rank compared to their peers based on performance. If you and the Captain are synced, cover their current enumeration based on performance and potential as well.
- Address their developmental needs (DA PAM 600-25 items, NCOES, advanced skill schools, SWCS tour, etc.)
- What is their motivation moving forward?
- What are their concerns regarding their career.
- Provide sustains and improves on current performance
- Review SM's goals from your last counseling session, did they achieve those?
- What went well / bad during this quarter? How can you do it better next time?
- What are your next quarter goals? How do we get there (plan of action)? Have SM help develop the plan of action.
- Review his Five-year NCOPD model (see figure 2:1) update with newly acquired schools, awards, language score, etc.

- DA PAM 600-25 and/or SF PDM review.
  - look like a commando, x 4 advanced schools, college, etc.
- Soldier Records Brief scrub (SRB), see paragraph 2-21 below for more information.
- Jump log scrub, ensure he is a current jumper. If a JM, is he current, and is he eligible for Senior or Master wings?
- DA photo review, once every 5 years, or after specific new award(s), and promotions (yes, still important even if not used during the evaluation boards).
- How is your immediate family, and how are your friends, parents, etc.?
- Impression of Special Forces.
  - And for your older NCOs: If you were the Team Sergeant what would you do differently?

2-14.3 Guys who are interested in earning the top spot on the team will ask the direct question "how do I earn the MQ block?" Have answers.

2-14.4 I strongly recommend including your seniors in counseling the junior guys. Depending on your comfortability they could handle (with or without your direct supervision) quarterly, positive and negative counseling sessions. As with anything, start small and have them cover little things such as a negative counseling on forgetting to tie down NODs. As you observe and mentor your Seniors, give them more responsibility in the counseling world. Doing this helps them prepare for their future as a Team Sergeant and gives them ownership as a senior on the team. Use as many opportunities as you can to push down the responsibility of train, coach, and mentor.

2-15 Not every Green Beret (GB) is a good fit for your ODA. You have a duty to train, teach, coach and mentor everyone assigned to your ODA. What makes this task difficult is every team has its own personality. When you identify an unresolvable conflict between a guy and the ODA's personality, you need to inform the SGM after you have exhausted all your resources.

2-15.1 If this individual is not a detriment to the Regiment (shows potential) but does not fit into your ODA's culture, do what you can to champion him to the other Team Sergeants in the Company. If you have done your due diligence with networking, you should know a good number of other Team Sergeants (in and outside your Company), and how they run their teams. This base of knowledge should help place the GB in question.

2-15.2 What if you are the 2nd or 3rd team he has landed on? This could be a simple oversite between his previous team's leadership and the SGM. What I mean is little effort was put forth to place this guy on a team where his personality would match the Team's culture. If this is the case, apply the above paragraph's method to help him land on the right team. Another potential option is that he just needs some time to mature. Depending on his MOS, a year on the B-TEAM, in the Battalion SIGCEN, etc. could be an option.

2-15.3 While we are on the topic of moving guys, be mindful if you suggest sending him to the B-TEAM, Battalion staff, Advanced Skills Company (ASC), or even out to SWCS. We need to send the right guys to these jobs. Think about yours, and the other team's products that could get mishandled if we only send our rejects to the staff. Think about the quality of new personnel we could receive if we use SWCS as a dumping ground.

2-15.4 What about the guy on the B-Team who was moved from a team due to a non-culture fit? Below, MSG Longoria provides a fantastic account of his experience with taking a chance on a guy and applying the teach, coach and mentor approach:

> (EL) Some guys may fall into the "problem child" classification due to a personality conflict with their previous leadership. It is imperative that you give some grace and an unbiased approach to your guys. Allow them to hang themselves with the latitude YOU give them.
>
> For example, when I took my Team, one of the guys assigned to me was under investigation for an incident overseas. This investigation lasted almost 2 years, and he was flagged, unable to attend any TDY schools or receive awards. When I met him for the first time, he was in language

school. His first comments to me were that he was under investigation for ___ and that he understood if I wanted to push him to the B Team. I told him that I had no intent to push him off and that he will have to prove to me that he wanted to be on my detachment, but he had my support in any way he needed until he proved otherwise. Eight months later we finally had his Flag removed. To this day, he has been the absolute best SSG that I have worked with in my military career. He ALWAYS gave me 100% effort, crushed absolutely every task I gave him, had a 2/2 in two Arabic dialects, took ownership of training concepts and completed without being asked.

When I left the Team, he thanked me for not getting rid of him and said that it was refreshing that I took a chance on him. Call it lucky, but this happened to me with two other Soldiers with a reputation of a troublemaker. At the end of the day, every ODA needs a pirate, you just have to be able to keep a tighter grip on those individuals with more direct guidance. Be prepared, this just may be one of the many leadership challenges you'll experience in your career.

To belabor MSG Longoria's statement **"every ODA needs a pirate,"** the legendary Marine Corps LTG Lewis "Chesty" Puller was quoted on this very subject... twice.

> Take me to the brig. I want to see the real Marines,[7] and If the Marine Corps expected you to behave, they wouldn't challenge you with a Good Conduct Medal.[8]

Obviously, I am not referring to the clowns who are 100% reckless and are always in trouble. Or the narcissist who cannot be bothered by those on the team. There is another version of the pirate, and I hope you are blessed with one of these rare breeds on your team. These are the guys who had the bad luck of being born in the wrong century. Labeled, "break glass in case of war" and they are generally not suited for garrison life. These men will never be a "yes man" in your formation.

Men like this can read the battlefield and anticipate what course of action the team needs to do. They are true warriors, respected leaders amongst the men, whose only concern is for the man to his left and right. He is there for one reason, the mission. You may spend a little extra time counseling him, but he is worth the effort.

It is possible that this individual will circulate from team time, to SWCS, back to a team, then to an Advanced Skills Company instructor slot. He may or may not see E8 and will never see E9. If you have done everything mentioned above, and this guy is not working out, do not beat yourself up. His personal life, personality, and other

---

[7] LTG (RET) Lewis (Chesty) Puller, date unknown, while on Battalion inspection.
[8] LTG (RET) Lewis (Chesty) Puller, date and location unknown.

factors are beyond your capabilities and resources to manage, let alone fix.

2-15.5 **Unfit for team life**. Sadly, you may end up with a guy who does not provide anything to the Team, Group, or the Regiment. Be prepared to create a Relief for Cause NCOER, bar him from re-enlistment, and possibly recommend a Tab Revocation. If the guy in question falls between a culture fit issue, and Relief for Cause/Tab Revocation there is another option, Change of MOS. Hit up your SGM to see which direction you need to go for any of the above.

I never had to do any of the above, so here are some pro tips given to me when I reached out to the network.

When you write a Relief for Cause NCOER, fight the urge to completely destroy the SM on the NCOER. Capture actions and/or attributes that are backed by Army regulation or policy:

- Physical fitness test failures
- Failed drug test
- Alcohol related incidents
- Domestic violence
- SHARP, EO, and EEO violations
- Dereliction of duty
- Missed movement

The why behind not destroying a guy was explained to me as such, most likely this is his first really bad NCOER. If all of a sudden, this guy who has never been on anybody's radar before, now gets crushed on one NCOER there will be questions. Or it can be seen as a one off, and the instinct will be to re-train and re-integrate him into team life.

I believe the instinct to crush a guy is due to previous leader(s) kicking the can down the road. Had they done their job this problem wouldn't be in your team room. Yup, it sucks, so what do you do? If you have identified a guy who provides nothing for the Regiment, it is on you to demonstrate a downward trend, or better said a pattern of failure or misconduct on paper. Sit down with your SGM, fill him in on the issues (preferably backed by Army Regulations, Policy, etc.), and ask him what specifically needs to be included in a packet of this kind. This will provide a road map for you to follow as you build up to this kind of NCOER.

The one thing I know you will be required to do is to capture every issue (anything that violates Army regulations, General orders, policy, etc.) on a DA 4856. Along with capturing the issue, develop a corrective action plan, and record the outcomes of those plans. Remember that some issues, such as an ACFT failure, comes with a specific check list to follow. For example, you cannot give the kid an ACFT and two days later give him another one... you must give the SM a certain amount of time in between events.

Do your homework, there is even a process for delivering a NCOER of this kind to a SM. Once your packet is complete, you will counsel the SM on his options after receiving one of these NCOERs. Again, I have never done this so adhere to the guidance given to you by your SGM.

2-16 Five-year NCOPD model (see figure 2:1-Five-year NCOPD model). The form is fairly self-explanatory and the importance of it is twofold. One, it highlights the SM's entire career in regards to DA Pam 600-25, and the Professional Development Model. They get to see a consolidated snapshot of how they look to a promotion board. Everything from deployment history, NCOES, recent assignment history, advanced schools, OML ranking, etc.

Two, the Senior rater ranking history, labeled "NCOER Review/5-year career plan," is a visual representation that benefits the SM, and the rater. Both can see the glide path of the SM, and the rater has a tool to structure counseling sessions. If your guy is like Rambo, John J. in figure 2:1, then you have your MQ guy or a contender at least. If his performance has not waivered, you can consider assigning tasks that give him more responsibility. This will set him up for promotion to E8.

2-16.1 If you have an average performer, both of you can see it. Review his past NCOERs, school history, etc. to identify where he got off track. Then develop a plan of action to get him back on the promotion path.

2-16.2 That being said, not everyone wants to climb the corporate ladder, and that is fine. These are generally fantastic team guys who work hard, but do not want, or do not show potential to take on the responsibility of leading the team. Do your part though, ensure he is getting to a new school every 1-2 years. Confirm that his NCOES is up to date, and the professional development model is being followed to the best of your ability.

Name: Rambo, John J.						Numeration: ## of ##						
Arrival:	30-Jul-20	MOS:	18B	DOR:	1-Sep-19							
PCS:	30-Jul-23	BASD/TIS:	08Aug2011-10yrs			DA Photo:	Mar-18					
RCP:	17-Feb-24	TIG:	2yrs			ERB:						
NCOES:	SLC	MRD:	N/A									
Language:	AD			College	4 yr degree, criminal justice							
Basic Skills: Airborne SCH, BLC, SFQC, ALC, AD SOLT, SERE-C												
Advanced Skills: SOTIC, MFF, SLIM												
Awards: AAM-3, ARCOM-2, JSCM, MSM, BSM, Parachutist badge, MFF badge, CIB, SF tab							Combat: AFG, LE x2, SY x2 /25 months					
	Position		SR 18B	SR 18B	SR 18B	SR 18B	SR Inst	Chief Inst	Chief Inst	TM SGT	TM SGT	TM SGT
	Rank		SSG	SSG(P)	SFC	SFC	SFC	SFC	SFC	MSG?	MSG?	MSG?
		NCOER Review						5 Year Career Plan				
	TIS	5	6	7	8	9	10	11	12	13	14	15
	Year	2016	2017	2018	2019	2020 COR	2021	2022	2023	2024	2025	2026
Most Qualified			X									
									Projected			
								X				Projected
Highly Qualified						X	X					
					X						Projected	
				X						Projected		
Qualified												
RCP SGT	14 years											
RCP SSG	20 years											
RCP SFC	24 years											
RCP MSG	26 years											
RCP SGM	30 years											
RCP CSM	30 years											

Figure 2:1 Five-year NCOPD model

2-17 NCOERs. If you are new to writing these, take the time to sit down with the SGM (early) and ask questions. He should be more than willing to teach, coach and mentor you on the finer details of the NCOER. If your Group has a NCOER Reference Guide, dig through it. These products are generally chalked full of great info. Even if you have a few reps in this arena, I still recommend sitting down with the SGM and digging into the organization's standards and nuances.

Once you know what is expected, start early on upcoming NCOERs. How early? Two months out was my sweet spot, here I would knock out a rough draft and walk away for about two weeks. During the two weeks I found myself thinking about the NCOER, and if something solid popped I would write it down. After the two weeks were up, I would sit back down and begin hammering out my final bullets. Starting this early provided enough time to go back to the SGM if I had questions, engage my peers, or research how to capture my guy's performance in the written word.

Submit early and turn in your best work. This shows the SGM you are dialed into your admin game, and that you care about your men.

Even if you believe you have turned in a literary masterpiece, be prepared for the red ink. Just accept the fact that your first few NCOERs will not meet the standard. This is part of the learning process, be humble and absorb what is being thrown your way. Another observation, every SGM/CSM will have specific items within the NCOER that are important to them. Most of these items are the new hotness coming down, which is a good thing for your boys. Stay tuned in for these changes, and don't lose your cool if you must make an abundance of corrections.

2-17.1 **Inspiration**. Don't sit and stare at an empty NCOER shell. Reference quarterly counseling's, training event AARs, hit up the local shared drive, portal, Army writer.com, the SGM, your Group's NCOER guide, and your previous NCOERs for inspiration.

I did what I could to capture NCOER worthy events in my green notebook during training events. If I was unable to obtain what I needed in real time, I would refer to the executed training concepts and AARs. Reading through them reminded me of who did what.

2-17.2 You should never have to write your own evaluation. This goes double for your guys. If you are stuck and cannot think of a bullet statement for "Leads" or have multiple empty blocks, ask your guy pointed questions. You know what training you have done for the last year. Bring up those training events, one by one, and conduct a play by play if needed until something of value comes up. You can lean on the Captain, 180, 18F for some input as well.

2-17.3 You must know how to write the Senior Rater (SR) comments. Learning how to write these requires sitting down with experienced E8s, SGMs, CSMs, Company Commanders, etc. Captains will lean on you, have answers for them. The Captain can also tap into the Company SGM and Company Commander for additional information.

2-17.4 You need to read the post evaluation board's field AARs, DA PAM 600-25, and review our career map (Professional Development Model) produced by HRC and/or SWCS. Knowing this information will help you mentor your guys. Show them the path they should strive to be on.

2-17.5 There is specific Senior Rater verbiage required to get your rock stars picked up on the Evaluation board. Senior Rater narrative comments that best support your Most Qualified Soldiers will change over time. Stay engaged, be sure to talk with your SGM, or other NCOER savvy Senior NCOs to learn what is hot. What worked during my time was as follows:

- Enumeration
- Strong recommendation for promotion
- Next required NCOES course, if complete, then an advanced school
- Potential in 1-2 years
- Potential in 3-4 years

For your Rockstar, what does the above look like written out?

- SFC Rambo is #1 of 4 SFCs I senior rate and is in the top 5 SFC's I have ever served with in 13 years of service. Promote to MSG now and select for SFOD-A Team Sergeant. Continue to groom for 1SG and send to MLC at the earliest opportunity.

How about your number two guy?

- SFC Jambo is #2 of 4 SFCs I senior rate and is among the top 8 SFCs I have served with in 12 years of service. Promote to MSG ahead of peers and assign as a Team Sergeant at the earliest opportunity. Send to MLC now.

Middle of the pack guys. These are either newly promoted E7s, or those who just love being a team guy. The following can work for your 3rd or 4th rated SFC:

- SFC Ruck-strap is #3 of 4 SFCs I senior rate. Promote to MSG with peers (or promote if slots available). Joe has demonstrated potential to serve as a Team Sergeant, continue to groom. Send to MLC.

2-17.6 For your number two and lower guys, be prepared to have those hard conversations. They may want to know why they are not number one if they are trying to follow that path. Here is where that hard work pays off. The quarterly counseling's will have captured what they did, or failed to do regarding professional development, MOS tasks, or overall performance compared to their peers.

Use DA PAM 600-25, or the SF Professional Development Model, and/or the Five-year NCOPD Model to highlight where they fell short.

Part of this discussion could include you informing him that his performance (up to this point) does not present the potential to cut it as a Team Sergeant. That he may be better suited as an ops sergeant, or an NCOIC.

Bottomline, your boys should not be surprised at NCOER signing time.

Pro Tip: Senior Rater comments is a go-to topic to bring up during those mandatory, awkward, high level CSM all calls.

2-17.7 You must understand the two Successive Assignments, and the single broadening Assignment blocks at the very bottom of the NCOER. The job titles should make sense for your SM's MOS, and actually exist within USASOC (use 18 Series Professional Development Model). Be sure to nest the successive and/or broadening assignments with the Senior Rater comments.

Without a USASOC Marketplace system (who knows, maybe one is in the works?), you must navigate the nebulous HRC website and find the list of jobs available to SF NCOs. There are POCs for CMF 18 representatives on the HRC website. I have reached out before with questions. Be careful here, asking questions is one thing, but trying to manage your guy's (or your own) PCS move is another.

2-17.8 Most Captains do not know how to write YOUR NCOER. Make sure the SGM and your Senior Rater are tracking your concerns (if any) with your NCOER. If it is substandard in relation to your performance, and potential, use your NCO Support Channel (big Army speak for SGM/CSM open door policy) to get outside assistance.

Warning: key words "outside assistance." If you and /or the team have any dirty laundry attached to your substandard NCOER, you may not want to bring that out. It may be time to look at the man in the mirror, and ask the hard question, do you deserve that lower rating?

If it is just laziness on your Rater and Senior Rater's part, fight like hell! No one cares about you, like you. Be prepared to write your own.

2-17.9 Just as listed above for your E7s, your Senior rater comments will follow the same sequence. If it is your second or third year in the seat[9] and you have been crushing it, your NCOER could look something like this:

- MSG Rambo is #1 of 7 MSGs I currently senior rate and the best Team Sergeant I have ever worked

[9] Before I was a team sergeant, 3-4 years in the seat could be expected for solid performers. When I was finished with my team time, Team Sergeants could expect the boot after two years. I am not sure if this trend will continue.

with in 16 years of service. Selected to serve as the next BSC 1SG, ready now to serve as an SF CO SGM. Send to USASMA at the first opportunity.

Or:

- MSG Jambo is #2 of 7 MSGs I senior rate and is among the top 5 MSGs I have ever served with in 15 years of service. Promote to SGM now and select to serve as a SF Company SGM (or-selected to serve as the next BSC 1SG, ready now to serve as an SF CO SGM). Ron has unlimited potential to serve as a CSM. Send to JSOFSEA at the earliest opportunity.

If you are finishing your first year as a Team Sergeant, do not expect a Far exceeds standards/MQ NCOER. Have an honest conversation with yourself, did you walk onto the job and have every facet of it dialed in? Did you perform better than your peers up and down the hall?

2-18 Identifying your top-rated NCO. Most teams have a few guys who are in the zone for the E7 evaluation board. Of those few guys you should have identified your number one, the guy you plan to give a MQ, and believe is ready to take the next step. Think of it this way, who do you want advising the Detachment Commander or the 180 while planning and running ops during split Team events? Who is that guy you can trust to run the Team when you are not around?

2-19 What do you do if you have a stacked team, how do you pick your number one?

My situation was pretty unique. Background: I was serving as a field team Chief Instructor in SUT within the SWCS machine. Although not an ODA, I had a majority of the MOSs and our job was rooted in mission planning and execution. Basically, we conducted four JCETs per year, six weeks apiece, training Soldiers with mixed experience to come together as a unified element and conduct combat patrolling.

On my field team I had three exceptional 18Bs, and all three were slotted as Primary Instructors. Each guy was a solid performer. They fully adopted the train, coach, and mentor mindset, and were lauded on every end of course critique by students. At first glance, they appeared to be tied for the position of E7 Chief Instructor (this title sets one NCO above his peers for promotions). All were mid-career (within a year or so of going indef.), averaged two combat deployments (three different operational Groups), similar awards, schools, team time, SFQC graduation dates, and so on.

With the guys being so close on the baseline data, I needed a black and white method to determine who was the most qualified individual for the MQ rating. The method I used, and in no way did I come up with it on my own (thanks SGM Delgado), was the following comparison tool. This tool takes most, if not all bias out of your decision.

Without another way to justify a decision, my choice could have been considered favoritism.

This method literally lists out what DA PAM 600-25 states regarding requisite skills, time in service, time in grade, etc. to be considered eligible for the next pay grade.

Once your comparison tool is built, use the SM's Soldiers Record Brief (SRB), NCOERs, fitness test scores, etc. to fill in the data, and tally up the numbers for a total score. If you utilized the five-year plan (refer to figure 2:1) in this chapter, most of the required data is there.

2-19.1 The comparison tool is simple to set up and use, which is good if you have been classified as a knuckle dragger like this guy. Below is the data I used. Yes, it is SWCS specific, but with a few changes it could better reflect an ODA member. As you can see, the top section is mostly baseline data with some score worthy items mixed in:

Legend:

+= Award point(s)

*= tie breaker point(s)

- MOS
- +Order of Merit (OML) number from the Army Career Tracker website (I was comparing 3 guys, whoever had the best OML received 1 point, no points for the other 2)
- +Time in Service (TIS)/Time in Grade (TIG)

    - be sure to reference the current evaluation board MILPER and/or 600-25 for minimum TIS/TIG required for the next pay grade
    - If this is the tie breaker point between top guys, you must look deeper... Not all old timers are ready to take a team
- ODA time
- *Time at SWCS
- +Completion of: Instructor Training Course (ITC), Battalion On-boarding, Master Trainer Course (All three courses were required in order to be considered a qualified instructor at the time of this book)
- +Have been awarded the Basic Instructor badge
- +Been recognized as Instructor of the cycle

Next, I listed out what 600-25 considers a Highly Qualified (HQ) E7 Career Management Field (CMF) 18 NCO being considered for promotion to E8. This list was reviewed to ensure each guy was worthy of further analysis. As you will see later on, the same categories are listed for the Most Qualified (MQ) list. <u>No points were awarded when reviewing this list</u>. A Highly qualified CMF 18 should have:

- 36 months ODA time
- SLJM
- Oral Proficiency Interview (OPI) or Defense Language Proficiency Test (DLPT) 1/1 or higher (this changed to 1+/1+ OCT 2022)
- Meets Height and weight

- Passing score on diagnostic ACFT (at the time the ACFT was not the official test of record, and USASOC had not published guidance on required scores)
- Current DA photo/updated SRB
- Commandants list for NCOES and MOS enhancing schools
- College, minimum 90 semester hours or an associate degree

According to 600-25 a <u>Most Qualified (MQ) E7 should have</u>:

- +12 months rated time as a Senior 18B, C, D, E or F with strong NCOERs with supporting enumerations (Far Exceeded Standards/Most Qualified), OR be rated as an 18Z
- +Strong Senior Rater (SR) comments which clearly state strong potential for advancement:
    - Promote now, promote ahead of peers, ready to serve as a Team Sergeant, future Sergeant Major
    - Successive Assignments: Chief Instructor (CI) SWCS, SFOD-A Team Sergeant

- +SLJM
    - Additional point if they are current and active
    - One could argue that Senior and Master rated JMs could receive additional points
- +OPI or DLPT 1+/1+ or higher
- +Meets Height and Weight
- +Consistently obtains 90 points or better, per event, in the ACFT

- +College, minimum 90 semester hours or an associate degree
  - Additional points awarded for each higher education level above associates degree
- +Distinguished Honor Graduate, Distinguished Leadership Award, or Honor Graduate, Commandants list from: NCOES or MOS enhancing courses
  - Each course with any of the above titles will receive 1 point
- *Valor awards (while not listed in 600-25, I also included MSMs and BSMs)
- +Looks like a Commando! Completed one or more advanced skill/functional courses: Ranger, Special Forces Sniper Course, SFARTAETC, Mountain (summer/winter), CDQC, MFF, Advanced Language training, ASOT, Special Warfare Training Course (SWTC), 18F course, Dive Supervisor, MFFJM
  - Each course listed above will receive 1 point each
  - This was briefed to all 18Zs multiple times by the SWCS CSM after each E7 Evaluation board (2020-2022), four schools seemed to be the magic number. Always a good topic to bring up when you are in an E8 and above all call with various CSMs
- +SWCS instructor (rated CI time?), Drill Sergeant, Recruiter, Dog handler, CI in an Advanced Skills Company (if no SWCS time, or an additional point if they have done both)
  - One point for each job position they executed

- Both SLC and Distance Learning Course (DLC) 4 complete
    - One point for both being completed, no point if DLC 4 is not complete... you would be surprised on the number of guys who do not have it done.
- *Current Jumper

Legend:

+= Award point(s), there are a few points to be awarded in the data section
*= tie breaker point(s)

With the data captured on each SM's sheet, total up each guy's points, and the highest score wins.

NOTE: What cannot be captured with a tool such as this is maturity, true operational experience, overall life experience, and someone that just gets it. You can only know this through quality interaction and committed observation. If you decide to use this specific, and subjective information for your decision, you will have to decide how and when to apply it.

Looking back to the NCOER section earlier in this chapter, you could use this comparison tool to highlight shortcomings of your second ranked and lower guys. It could help you with those hard conversations when they want to know where they fell short.

2-19.2 Enumeration. Keeping your guys racked and stacked requires consistent effort. When I was on top of my game, I used a printout of my NCOER Tracker, E7s grouped together, same for the E6s. Next to their name I would pencil in my rating based on performance, and then I would engage the Captain for his assessment on potential.

Again, the difficult part of this process is reducing bias. There is no way to completely remove bias, we are human. For example, when one of your guys gets back from a mission critical advanced school, and at the same time, another guy completes his Batchelor's degree. If these two are your top performers, who gets the MQ? I don't have that answer. You and the Captain will have to drill down and conduct a whole person evaluation.

On top of the above dilemma, your rankings must match, or be damn close to the Captain's. Example: If you rate a guy's performance as a 1 of 4, and the Captain Senior Rates his potential as a 4 of 4, HRC is going to kick it back. Or that NCOER will be ignored during evaluation boards.

2-19.3 **At the end of the day, you (and the Captain) owe the guys three things with NCOERs.**

<u>One</u>, where they stand on the team, and NOT to be shocked with their enumeration at NCOER signature time.

<u>Two</u>, a fair and honest evaluation of their performance, and potential.

Three, a timely evaluation. Most likely, your E6 and E7s are go-getters. They are probably 100% focused on their job, the team, and family. You must be aware of the evaluation process, and the specific time of year each board convenes.

Nothing worse than getting a text from one of your former guys, after the board, and he informs you that his NCOER did not make it on time... I did not prioritize his eval during my PCS. When you have guys that bend over backwards to ensure the mission is accomplished day in, day out, it is gut wrenching to let them down. As a result, he was not picked up for E8 that year, even though he met all of the requirements to be extremely competitive.

If you have a solid pipe hitter (Far exceeded standard and Most Qualified), and his annual NCOER is not due until after the evaluation board, talk to your SGM about possibly doing a Complete the Record NCOER. The board will always publish a MILPER message with instructions covering this topic. Be sure to follow the directions carefully. If his annual NCOER due date is aligned with the upcoming board, ensure to get the evaluation done, and S1 submits it on time.

One more time, learn the Evaluation Board and promotion process. You are the dude who is responsible for the professional development of your guys.

2-20 Passing the torch. You need to teach your guys how to write an NCOER. A good place to start, but not everyone's favorite, is the NCOER Support Form. This form does a few things, first they learn the admin data section. Next, they set goals, and (with any luck) as they achieve them, they see how those become bullet worthy events. Their goals should be based on professional development, the ODA's upcoming mission, and anchored in the ARSOF attributes.

Not every rater utilizes the NCOER Support Form, I too, am guilty of this. I started strong with this method when it first became available on EES, but I did not prioritize it as the years went by. Looking back, I missed a huge opportunity to help my senior guys develop as future leaders.

Just as I suggested earlier, if you believe your seniors are ready - they have demonstrated a certain level of proficiency with standard counseling – have them help the junior NCOs create and fill out the support forms. Teach them how to find the data required and how to fill in the admin portion of the support form. Show them how to arrange the bullets to capture what you want to say, while keeping them in the ridiculous two-line format. Above all, they begin to see how to quantify a SM's performance.

2-21 SRB. An important part of the Evaluation board is the SRB. Below I covered what to look for in your capacity as a Team Sergeant. Covering all the details regarding the record brief is outside the scope of this book. If you want to learn more, sit down with the S1 NCOIC, or execute an internet search.

2-21.1 Reviewing this document with your guys does two things. One, you ensure completeness for evaluation boards. Two, it serves as a NCOPD when discrepancies are discussed with your SM.

Section I Assignment Information:

- Overseas (OS)/Deployment Combat Duty, are all his trips captured?
- Dwell time, does it align with his last listed OS/Deployment Combat duty time?
- Is his MOS information correct for PMOS/SMOS/Bonus MOS?
- Does his SQI, ASI info match schools attended?
- Flags, does he have one? If yes, does he know about it? Every flag code brings a different course of action to get it lifted. Consult with your S1 to learn what needs to be done.
    - The Career Counselor is a great source of information regarding Flags.

Section II Security data:

- Is his security clearance info correct? You are looking for type of clearance, and currency date

Section III service data:

- Are the BASD, ETS, etc. dates correct?

- Army Good Conduct Medal (AGCM) date and eligibility date, correct?
  - **Pro tip**, Number of good conduct medals should match the number of service strips on the Army Service Uniform
  - This category becomes difficult when dealing with guys who were National Guard or Reservists prior. Consult S1 if issues present themselves.
- Date of Rank, does his current rank and DOR match what is listed at the top of the SRB? If not, this could be an issue when he goes to ETS/retire

Section IV Personal/Family data:

- Is the personal data correct?
- Is the physical up to date?
- Medical readiness block, MRC, can he deploy?
- Height/weight, ACFT (APFT), I was told (back in the day) that this data needed to be within the last six months prior to an evaluation board
- Is the home of record and current mailing address, correct?
- Dual military families come with extra work. If they have kids, is the family care plan current, and filed with the appropriate offices/personnel? Is the spouse's info listed?

Section V Foreign Language:

- Is his assigned language, and last test date listed?

Section VI Military Education:

- Military Education Level (MEL), Military Education Status (MES), does this match his last NCOES course completed? Has the SM completed the requisite Distance learning module for that NCOES?
- Military courses, are they all listed? If they were the distinguished honor graduate or made the commandants list, is this captured in the Achievement column (ACH)?
- Weapons qualification (BMQ) data listed and current? S1 pulls this data from DTMS

Section VII Civilian Education:

- Highschool or GED info listed?
- Higher-level education institution/discipline information listed?
- If a higher-level degree has not been completed, are the semester hours listed? S1 will need an official transcript from the college(s) in order to add the hours

Section VIII Awards and decorations:

- Are all awards, tabs, badges, etc. listed?
- Does the list of awards match the DA photo?

Section IX Assignment Information

- After a few conversations with S1, and super savvy SGMs, opinions differ on this section. I agree that in the Duty column each deployment should be listed, and be titled FWD. This paints a picture for the evaluation board on the SM's pattern of life in conjunction with Section I. For the most up to date info, hit up your S1 and Sergeants Major.

Again, this is not a complete list for all things regarding the SRB. I tried to be generic in my scrub info because S1s will differ slightly from Group to Group.

2-21.2 Think of scrubbing SRBs this way, the evaluation boards only get so much time with each packet. If the Service Member's SRB is squared away, that could be the impression the board has about the Soldier. Better said, if the Soldier can't square away his own SRB, a basic individual task, why should we trust him to take care of other Soldiers?

2-22 Sending your guys to SWCS. Some GBs will escape the gravitational pull of SWCS, many will not. When levy season arrives at Group, be proactive. You should know who (on your team) is in the window for possible assignment. Always ask for volunteers, you might be surprised at who says, "send me." Lastly, know who needs to take a knee, 99% of the time they will never say it out loud.

For those going against their will, or the others who volunteer, talk to your guys about options. Everyone has different interests or skill sets and in most cases; Mountain guys want to be assigned to the schoolhouse in Colorado, Sky gods want to go to Yuma, and long-range shooting addicts desire SFSC.

Looking back to Chapter 1, two general rules come to mind for this topic. One, don't make promises you cannot keep and two, you cannot make everyone happy. No matter how hard you try, not everyone will get their primary committee choice.

2-22.1 If you are grooming a future Team Sergeant, who is facing SWCS assignment, I would suggest either SUT, or Robin Sage. These, in my opinion are the best committees to prepare an E7 for the Team Sergeant job. Specific professional development perks associated with SUT are as follows:

- SUT resets the mind back to the basics. If you want to do the 18Z job, you have to be the reality check for the 18Bs. You do not have to be a master tactician to be a Team Sergeant, but you have to know the principles of patrolling in order to keep the team grounded. You must ensure that the ODA is not overlooking fundamentals that keep men alive.
- Management of up to three other cadre members
    - Course prep
        - Coach, teach, and mentor new cadre on class presentation
        - Inventory, request, and receive course material and equipment
        - Requesting, securing and inspecting training locations
        - Rehearse and refine lesson plans

    - Course conduct
        - Issue and maintain accountability of course sensitive items and equipment
        - Manages troops to task calendar
            - ensures assigned cadre, and students are prepared for training
            - Ensures cadre/students arrive at the right place, right time, right uniform
        - Coach, teach, and mentor up to 18 students
        - Manages training ammunition, and pyrotechnics
        - Countless hours perfecting the art of counseling (written/verbal), to include counseling Captains
    - If assigned as the E7 Chief Instructor, it is an opportunity to serve as, and be rated as a Master Sergeant
- In-between course perks
    - The current course frequency (4 course cycles per year/6 weeks each) provides ample time to attend Military schools (and college) in-between classes
    - 3rd SFG consistently runs USASOC SLJM courses, and is easy to get school slots
    - Maintain Jumpmaster currency/work towards Senior or Master ratings (Static line + MFF)

- Each MOS committee is open to cadre shadowing course material in order to re-fresh
- Ability to earn the Basic, and possibly the Senior Army Instructor Badge

2-22.2 At the end of the day, don't dread the SWCS levy. More than likely, you have done a tour. Share your story with the guys who are eligible, sell the good parts, and steer them from the less desirable jobs. Whether you have volunteers, or it is just a guy's time to go, engage with your SGM to see what jobs are available. Do what you can, and fight to get your guy(s) lined up for their preferred job, or one that best suits them.

2-22.3 If the trend of promoting younger guys (10-15 years' time in service) to E8 continues, the following should remain applicable... If you are in your last year as a Team Sergeant and you are looking at 3, 4, or 5 years before you are eligible for retirement, I would suggest volunteering for a Chief Instructor role at SUT. Why not come share your 15 years of SF knowledge with mid-career E7s? Also, you have hands on access to Captains coming through the course. I cannot think of a better place to have served my last few years in the Army.

2-23 Re-Up. Know when your guys are up for re-enlistment. If it is your guy's first re-up, or even if it is his third, go with him to the career counselor. You can help him ask important questions, and ensure the counselor isn't greedy with the Swag. Regardless if your re-up experiences were good or bad, you should know of the potential issues that can arise, and help your guys navigate theirs.

If your upcoming mission is aligned with someone's re-enlistment window, hit up the S1 to see if the location is tax free.

2-24 Awards. Every Group seems to have a different understanding of AR 600-8-22. A trend has appeared, not just in our Regiment, but across the Army. Rank seems to have become a pre-requisite for certain awards.

With that said, your job is to ensure the hard work your guys did is properly recognized. Although, writing awards in my opinion, is on the same interest level as Drill and Ceremony in our Regiment, it is 100% in our (NCOs) lane. Again, reaching back to the NCO Creed, "**I will be fair and impartial when recommending both rewards and punishment.**"

I am guilty of pushing this responsibility onto my Captains. We are the guys who are with the men day in, and day out. Why wouldn't we be responsible for this task? Absolutely, the Captain should be involved in the awards process. Let him put that college training to good use and review for sentence structure, voice, a thesaurus and spell check, then send it up.

2-24.1 How to get started, I suggest reading the regulation. Determine which award is congruent with the effort put out by your guy and submit it. It is the job of the approving chain to endorse the recommendation for approval, downgrade, or disapproval.[10] You should not have to conduct a re-write to reflect a lower award. If this is happening in your Battalion and or Group, hold onto this little knowledge nugget for a later date. If you decide to stay in and climb the corporate ladder, one day you can be the voice of reason with the boss. Rant complete.

Use common sense when selecting which award to submit. Work within the boundaries of reality here. If you submit your junior 18C for an MSM, and his write-up summarizes that he served 36 months on your ODA, and his biggest contribution was zero equipment discrepancies during change of command inventories... be prepared for disappointment.

When you look up the MSM in 600-8-22, the overall definition alludes to having affects at the organizational and even strategic levels.

---

[10] AR 600-8-22 (5MAR2019) 3-21, k. All requests that are not processed while the Soldier was assigned to the organization or in theater are considered retroactive, and must be processed through the former peacetime and/or wartime chain of command which was in effect at the time of the service or achievement to be recognized. Chain of command is defined as the sequence of Commanders in an organization who have direct authority and primary responsibility for accomplishing the assigned unit (cont.) mission while caring for personnel and property in their charge. Commanders in the former chain of command, (for example, BN, brigade, division, Corps, and so forth), to include the awards approval authority for the request, must endorse the recommendation for approval, downgrade, or disapproval as appropriate in the intermediate authority blocks on the award form. Every attempt will be made by the recommender to obtain the original chain of command's endorsement for all award recommendations. In the event that a member of the former chain of command is not available, the recommender must provide documentation verifying they have taken all reasonable steps to locate the appropriate official(s).

2-24.2 Just like NCOERs, don't stare at a blank award form. Review his last few NCOERs to see what he did during his time on the Team. Don't be ashamed to dig through the shared drive for inspiration.

2-24.3 Have your 18D ask for example award shells from S1. The backside of a DA Form 638 can be confusing. The example shells should have the current approving chain data filled in, along with other key items. Have him learn the whole form and teach other senior guys how to fill it out. There is always time for Professional Development.

Last thought on awards, if you wrote one or more impact awards for a guy while he was assigned to the team, you cannot use those events on his PCS award.

3

On Training

"Luck is what happens when preparation meets opportunity." –Seneca

Where to start?

3-1 Thanks to our core missions, team construct, and plethora of advanced skill sets, the ODA can be sent almost anywhere in the world to tackle a variety of complex problems. These facts bring up big questions when looking at a blank Long Range Training Calendar (LRTC):

- Can we achieve "T" for trained on all of the core mission sets? If not, which one(s) do we focus on?
- If my Team's Specialty infill method does not align with the projected mission assignments, how much time do I need to dedicate to it?
- What resources do I have at my disposal to create a worthy training plan?
- How and when do I schedule schools for my guys?
- What is my ODA's budget for the year?

The goal of this chapter is to answer these questions and a few more.

3-2 When I was handed the keys to the team, the first six months were filled in from the previous guy. Shortly after taking the reins though, Battalion wanted to see our next twelve months filled in on our LRTC.

3-2.1 As a new Team Sergeant, your story should be the same, but in life, timing is everything. You may be walking into a situation where the team has recently imploded and one, or all of the leadership has been fired. In this scenario, I would confirm with the Company Command Team if the ODA is on track to execute its original mission. Or due to whatever happened, has that mission been given to someone else?

If the team's original mission has been canceled, or given to another team, you now have the ability to build the team to your liking.

If the team is expected to continue mission, here are some items to focus on:

- What is the mission? (OCONUS event)
- When does the team deploy?
- What training phase is the team in? (Individual, Collective, FMP)
    - If the team is in the FMP phase, check DTMS and see if your late arrival meets the leadership requirements for validation
- What training does the team have planned during each phase?

- Are training concepts for each phase of training approved?
    - Have training resources been secured for all concepts?
- How many of those concepts are not approved, and where are they in the staffing process?
    - At the team level (returned to ODA for corrections), or at the Company/ Battalion/ etc.?
    - Not complete, waiting on medical plan?
- Does your infill specialty apply to your upcoming mission?
    - If yes, are the training concepts created? Are the concepts designed to build upon one another (crawl, walk, run) then onto a certification event?
    - If no, how much time do we dedicate to the skill IOT maintain a minimum proficiency?

With this information, you have enough to develop a reverse timeline, prioritize, and execute.

3-2.2 Regardless of the scenario you fall in on, the Marlow method is tried and true, see figure 3-1. SFODA 12–18-month planning cycle.

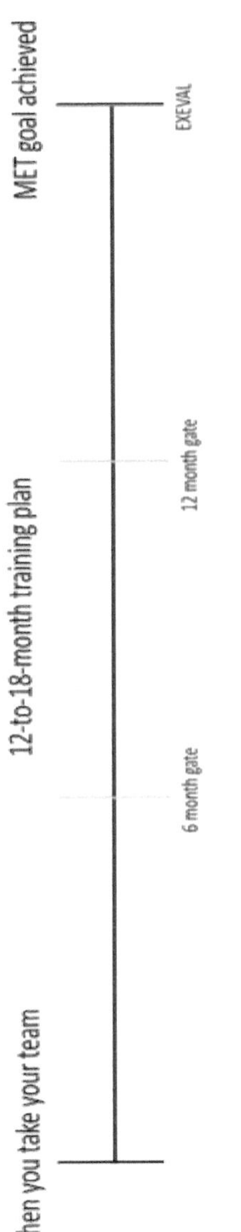

- Review/understand annual training guidance, your ODA's projected AOR(s)/named Operations/TSCP
- Top three will discuss the training direction and what METs will need to be focused leading up to the mission
- Top three will talk about what gates need to be in place to measure the glide path of training
- 180A will lay out the long-range training calendar (Everything from leave to load out)
- 18Z will start planning the training up to the 6-month gate
- The 6-month gate will show if you are on track as a team or need to re-train in a certain area
- The next 6-month training block will build on the initial 6-month training block if the gate is passed
- This will continue until you reach the 12 or 18-month mark will you will arrange an external evaluation
- This is the easiest and best practice to set up substantial training for your team
- Utilize CATS (Combined Arms Training Strategies) to understand your tasks and sub-tasks for each MET.
- All training should start with the fundamentals and end with an FMP utilizing as many MOSs as possible

Figure 3-1. SFODA 12–18-month planning cycle (The Marlow Method)

LRTC tip number one, Learn the Commander's Intent

3-3 Before you start building out your calendar, get a hold of your Battalion Commander's Annual Training Guidance. This document should be constructed to fully support your AOR's TSOC's Mission Guidance Letter. Think of this document like the Commander's Intent within an Operations Order. The guidance is the sum of the strategic vision from higher, current intelligence, and predictive analysis within your Group's Area of Responsibility. This summation will spell out the training priorities (core mission sets) for the ODAs to focus on. The Annual Training Guidance should be available on the portal and/or the shared drive (S3/S5 would be a good place to start).

3-3.1 **What is on deck?** Next, sit down with the Future plans' office (S5 / S35) and look at the Battalion Horse blanket (LRTC). If your ODA is aligned with a named operation, or a Theater Security Cooperation Program (TSCP), then start building your calendar to prepare for that trip.

3-3.2 If for whatever reason your ODA is not bound to either of these, or any variant of an OCONUS mission, you might be on a re-building year. Nothing to worry about, if you follow the planning principles in this book, your ODA will be noticed by the boss in a good way. A re-building year is a good time to pair training to meet the BC's training guidance and run full steam into a level one proficiency for your infill specialty.

3-3.3 **Research your mission**, put that TS clearance to use and spend time digging through OPLANs. Focus on your assigned mission, know how it ties into the OPLAN.

3-3.4 **Mission Essential Tasks (METs)**. Now armed with a focused direction, the top three discuss which METs the team will need to accomplish to best prepare for the upcoming mission.

3-3.5 **Build your LRTC**. After the METs have been decided on, depending on your mission, fill in known hard dates (or your best guess based on past events) on the LRTC. Examples: PDSS, ADVON, Load-out, Main body departure, and return dates.

3-3.6 With your mission's hard dates in place, divide your calendar into three chunks of time, or better said, three training phases. Generally, the first two phases are six months in duration, and the third can vary depending on complexity, training location(s), and resources. The blocks of time may be shorter depending on the deployment's departure date.

LRTC by phase, each phase has a purpose:

- Individual tasks (Soldier, MOS tasks, and Advanced skill schools)
- Collective tasks (Shoot, move, communicate, medicate, and survive thrive as a team)

- Full Mission Profile (FMP)/Validation

Each phase should have a defined end state, or as the Marlow product (Figure 3-1) states, "a gate." Something measurable that indicates that the team is ready to move to the next phase. For example, the ODA can move onto Collective training when we have fill in the blank number of MFF JMs, our 18D returns from SOCMSSC, and our JTAC is current.

3-3.7 What if your team is full of young guys? Depending on their maturity, and how quick they adapt to the training, be prepared to maximize your time in the first two phases. Maybe your team is full of seasoned E7s, and the detachment runs like a well-oiled machine.

If your situation is the latter, you may be able to throttle back on the collective training phase and dedicate more time to individual training. Or should more time be dedicated to collective tasks that have been put on the back burner... like MDMP, your infill specialty, team trigger time that does not include a flat range?

Regardless of the two examples above, or the dozens of possible variations, the team's leadership gets to decide how much time is spent in each phase. It is rare for the Battalion or Company leadership to reach down, and dictate the how, to an ODA.

3-3.8 **The Team Sergeant's roll**. After the METs have been decided on, and gates have been agreed to, this is where your experience comes in. You know how to link the medley of Soldier skills, occupational specialties, and Advanced skills (individual skills phase/first six months), into collective tasks. You know that Collective tasks (second six months) start small, and continuously build on one another until the team is ready for validation. To conclude these training cycles, you know how to develop a full mission profile/validation event that truly test the team for the mission.

3-3.9 **Who owns the LRTC?** If you have a 180, he is responsible for developing and managing the team's LRTC (everything from leave to load out). If not, you and the Captain will need to tackle this product. Before you get ahead of yourself, fill in Federal and USASOC holidays. As you should know by now, you will never participate in all those holidays. Take note of those missed weekends and work them into your 6–9-week calendars. Your leadership should be cool with this. Just be upfront with your SGM when he reviews your training calendars.

LRTC planning process considerations.

3-4 Before we deep dive into the three phases of the LRTC, keep the following in mind throughout your planning process.

3-4.1 Fact, you cannot be "T" for trained in every core mission set within Special Forces. There are not enough training days in the calendar year.

Example: I had an amazing Battalion Command Sergeant Major at the time. I was young and wanted the ODA to dominate. We had a LRTC that encompassed damn near all of the core mission sets to include the use of enablers, and we incorporated our infill specialty into most of the training.

At the end of the Semi-Annual Training brief, the CSM pulled me into his office. His direct question to me was, "Where is the time off for the boys?" He was concerned that although my team was on track to knock it out of the park, he saw the inevitable break down of the troops. He was worried that the guys would reach tracer burnout and make some epic rockstar level mistakes. Which of course, would ultimately derail the team from the mission.

He helped me comb through the calendar, keep the most relevant training, reduce redundant events, work in admin weeks, and more time off into the schedule. He helped me see through some of the erroneous priorities, shifted my focus back onto the strategic goals of the region, and how to dial down the breadth of our LRTC.

3-4.2 ***Leadership Principle 1, Be technically and tactically proficient.*** Always prioritize the basics. Simple Soldier tasks, individual MOS skills, MOS cross training, shoot, move, communicate, and medicate, are the foundation to thrive as a team. These are the fundamentals to success in every core mission we have. How can we justify spending time, money, and resources on "advanced" training, if our guy's don't know the ODA's hand and arm signals, or how to execute battle drill 1-A?

3-4.3 **Ammunition**. Bullets are pretty clutch for training. If you were not an 18B, worked on the B-TEAM or in the Battalion S3 as the ammo NCO, then you might not know the following. Every year each team is required to build out an ammo forecast. Each Group is allotted a certain amount of ammunition per year. This number is divided down to the Battalions and companies either in equal amounts or is given to the most aggressive command teams. Knowing how your Group handles this can save you a headache down the road. I would hate to do weeks of concept development, then come to find out half the range work you wanted to do cannot be supported.

Within the first month of you getting there, while you are digging through the upcoming training concepts, take the time to deep dive the previous Team Sergeant's ammunition forecast. You are looking to identify planned ammunition numbers in the training concepts, and if the Company is tracking these requests. The outgoing guy's numbers might be spot on with the LRTC he left behind... or not. This might require an un-forecasted ammunition request. I am sure each Group attacks this differently, talk to your Company and Battalion ammo NCOs, and learn how the process is done.

Ask to see the STRACK ammo allotments. If you understand how the STRACK ammunition works, you will have a stronger case when arguing for your fair share. Also, find out when your Battalion normally requests the next year's ammunition numbers, and add that to your yearly battle rhythm.

NOTE: Your 18Bs should make friends with both the Ammo NCO and the Armorer.

3-4.4 **Airborne Operations**. Jumps, yes, we still test gravity. Don't be the Team that just shows up for the Battalion jumps to maintain currency, although, sometimes unavoidable. If possible, you should always:

- Get manifested as a Team, same stick
- Plan for and use an assembly area (not at the chute turn in truck)
    - Or plan for split assembly area so you can conduct link-up (near, far/day/night recognition, running password, number combo, etc.)
- Prepare bump plans
- If unable to get the whole Team on one pass, treat it like a split Team event
    - Cross load leadership and equipment. Make your 18C do his job with the equipment cross load plan and weighing rucks.

Use jumps in your FMP events if you can. Requesting your own air takes some effort but the juice is worth the squeeze. Otherwise work with the B-TEAM to request the air, so you are the main effort, and the B-TEAM can do the heavy lifting while running a Company (which always turns into a Battalion wide) currency jump.

Below is the method I tried to follow when planning jumps for the team. I have never been a MFF guy but I am sure these principles apply:

1. Day/night slick (during individual training phase Advanced schools/MOS).

2. Day combat equipment (during individual/MOS cross training/collective training phase).

3. Night combat equipment (during collective training/ FMP phase).

As a Jumpmaster, you should make the effort to know as much as you can with this skill set. Many arguments can be made to why this infill method is considered antiquated, but until Starfleet command drops this task, we must continue to train for the possibility of using it.

Dig into USASOC 350-2 prior to Airborne Operations. Read the USASOC safety messages that come down. Manuals will be updated frequently, stay on top of this. A resource new to me is the Airborne Operations Detachment (AOD). They can provide you the latest and greatest for all things SLJM and MFF.

Know your installation's Drop Zones, and the yearly weather trends. Make friends with your BN S3 Air NCO and riggers.

Make friends (if you haven't already) with the JMs you see at every jump. These JMs are generally a wealth of information when it comes to running jumps.

Build and use bundles, your 18C needs the reps. Riggers are always looking for work other than packing chutes. They can help your team re-blue on bundles.

If you have made friends with the riggers, you can hit them up for JM refresher training if out of currency.

3-4.5 **Team Leave**. Build in two block leave times, Summer and Christmas are standard. Sometimes deployments will preclude the above block times from being used. Make sure to schedule time off before and after trips.

3-4.6 **Bulk Administrative requirements**. Set aside time, November through early December worked best for me, to knock out as much of the annual online training as possible. Also, I scheduled SRP around this time. SRP is normally done once a year if no deployments are scheduled.

> (EL) In 5th SFG, SRP packets were often an inspectable item during QTB's. The Battalion CSM would often pull packets from the BN S-1 and check to see if all ODA members had the required information. To sell the Team, I ensured that all SRP packets were 100% identical to one another, with regards to contents, and the order

the documents were placed in the folder. This may be a little excessive, but this is a good project for a young 18D, and it may push your Team over the hump to get a solid mission, if this were a deciding factor.

3-4.7 **Battalion level training briefs**. Every Group is different, but the team will brief the boss quarterly or semi-annually. Each Commander will run these briefs in a particular manner... some want the whole ODA to brief, and others only want to hear the Detachment Commander and Team Sergeant.

The intent of these briefs is to convince the Battalion Commander to invest his progressively shrinking budget on your projected training concepts. Take it to heart the numerous times this book suggests tying your training path, to the Commander's Annual Training Guidance. This will increase the chance of your concepts being approved.

Be well rehearsed, demonstrate your Team is synced, and running on all eight cylinders. The Captain must be on point to execute his most important duty: market your team up and out. He needs to do this without kissing too much ass. The sales pitch needs to reveal how the Captain has implemented the boss' Annual Training Guidance into his own Commander's intent. Lastly, Rehearse, Rehearse, Rehearse.

3-4.8 **LRTC construction frequency**. Planning out your LRTC will happen a few times while you are in the seat. As I mentioned in Chapter 1, you will need to engage in this task as soon as six months into your tenure. LRTCs normally "expire" after your assigned mission concludes, this means the top three will need to plan out another LRTC before you head out on your trip. This is due to the lead time to request resources, such as an ammo forecast. Additional reasons to develop a LRTC: starting a re-build year, you have a large turnover in personnel, or the Annual Training Guidance has changed dramatically.

Where do you start? What is your new mission compared to your last mission? If the new mission is a different core task, then determine the required METS for the new mission and build your LRTC from the basics to complex tasks that support those METs. If you are executing the same core task/METs, review your lessons learned and build out your calendar.

Pro tip: Check with Battalion to see how long core mission validations last. How long does your team's "T" for trained stand? When does it fall to "P" or "U"? If you have not done so already, or it has been a while, brush up on your doctrine knowledge before heading up to Battalion. Read FM 7-0, Chapter 1, paragraph 1-11 (figure 1-1) page 1-3 before you engage with the staff.

Just like your initial assessment period (Chapter 1), evaluate the team. This time around should feel different because you now know the team. Be careful not to assume that your team can execute a task, ensure they can do it by taking the time to test them. Again, start with the inexpensive route and quiz them. Look to the skill level 1-4 books, CMF 18 skill level 3 and 4 manuals, MDMP references, Ranger handbook, and Core mission set manuals. Develop your 10-25 question quizzes to determine a starting point. Different from your initial assessment period, you now control the LRTC. You could front load a physical/hands-on assessment of many tasks before building out an extensive LRTC. If the guys are able to execute tasks to standard, this could open up more time to dive deeper into more complex tasks related to the new mission.

3-4.9 **Funding**. Not every Commander is the same when it comes to money. Some want to spend every cent they can on the boys and fight for more. Others think their OER is directly tied to how much money they can save the unit. The Regiment's and unit budgets will grow and shrink with the ever-changing world situation. Be prepared to create low cost, high impact training for the guys... focus on the basics.

Growing up on an ODA within the GWOT money years was fantastic! If we could link a new commercial off the shelf training venue to our current mission, it was resourced. As I transitioned into the Team Sergeant roll, the cutbacks began. Our justifications had to be iron clad to convince the boss to open his wallet.

A big chunk of the boss' budget goes to training and certifying the specialty infill teams. The number of certified specialty teams in the Regiment will flex, and every Group maintains a different number of specialty ODAs. One reason for this is the vast difference in geography of each AOR. For example, 10th Group dedicates money to its Mountain capability when 5th Group historically does not.

Then there is the rest of the budget. The better your ODA is at nesting your training directly to the Commander's Annual Training Guidance, TSOC mission letters, or a named operation, the bigger chunk of cash you could get.

Another avenue to the Boss' wallet is the Battalion XO. Chances are that you personally know him, worked with him, or can identify an obscure connection between the two of you. Whatever the connection is, lean on it. He will know which pots of money are available, or when they will be freed up. Like any relationship we establish, be sure to take care of him.

Lastly, CSM Dorsh recommends the following:

> You and the Captain need to learn the concept of "Other People's Money" (OPM). CTCs (NTC/JRTC) have a lot of money associated with those rotations. An easy way to get training and equipment paid for is to sign up for a rotation. Painful, but can serve as a funding source and a

Certification, Verification and Validation (CV2) pathway for a deployment.

3-4.10 **Battalion's playbook**. The 180, Captain, and you should visit the future plans office (S35 or S5) quarterly to make sure the unit LRTC hasn't changed. Do not be surprised when changes happen, they can be drastic and/or violent to your training plan. When this happens do not lose your mind, confirm the changes through your chain of command. If the change is legit, do not throw your LRTC into the trash. Follow the same steps listed throughout this book to see if some or most of your planned and resourced training will work for your new mission.

3-4.11 **Condensed timeline**. Sometimes a mission comes down the pipe and needs an immediate fill. I have not personally been through this ringer, but I have observed a few ODAs jumping through hoops to get out the door.

In these scenarios Commanders will have to assume some, or a lot of risk. The level of risk will be determined on training and certification corners that must be cut, to meet the deployment timeline. Understanding the Commander's intent (GCC/TSOCs) will help guide you and the Captain towards the tasks required to successfully accomplish the mission.

Below is a short list of risks Commanders will need to make a decision on. Obviously, there are countless more, this is enough to get the discussion started.

- Required advanced skills for this mission
    - Is there enough time to send one of your own to a school/refresher training? No, who can you borrow in the Company, Battalion, or Group?
    - Bringing in an outsider is a risk; trust takes a hot minute to build
- Required specific MOS skill set... 18D, if he is not current, is there time to send to SOCMSSC? If not, will the Commander/Battalion surgeon accept this risk? Can the Commander/Battalion surgeon assume this risk anymore? Your 18D's credentials could expire when deployed, will the surgeon accept this risk?
- Collective training and FMP, depending on the timeline there may not be any available resources, ranges, ammunition... to conduct training. Will the Commander accept the risk of skipping a validation event?

3-4.12 **Time to grind**. Now that the leadership trio understands the strategic importance of your upcoming event, you have taken into consideration what core mission set(s) the boss wants your team to utilize, and your calendar is divided up, it is time to plan the next twelve to eighteen months.

Training using a phased approach: Phase I, Individual skill phase: Soldier skills, MOS tasks, and Advanced Skill schools.

3-5 This is the time to not only build up your team's capability for a specific mission, but it is also a good opportunity to help develop your guys for their future. As I mentioned earlier, "look like a commando" is a thing in order to get promoted. If time permits, send those deserving guys to as many schools as they can attend. You will need to be the voice of reason, send them to NCOES (including the corresponding DLC), Static Line JM, MFF JM, Dive SUP, Mountain leader, etc. After they have these schools, then get them into the fun courses.

3-5.1 Push them to enroll in college or finish up that degree they started five years ago. College has become the new norm for NCOs. The NCO evaluation board process has made it clear, guys with 4-year degrees have been receiving better OML numbers.

The school issue (miltary or civilian) will be a balance between what skills your guys need for promotion, and what the team needs to accomplish the mission.

3-5.2 Advanced schools, if you do not know where to start, dig into 1st SFC(A) Regulation 350-1. Chapter 5, 5-14, table 5-1 has a list of the minimum advanced skill requirements per ODA. You can also visit the Force Management System website (FMS WEB). It is the site that lists every Modified Table of Organization & Equipment (MTOE) for each organization in the Army. If you are still lost, sit down with the SGM.

3-5.3 **Detailed break-down of the school's process by Enrique Longoria**:

(EL) With experience and knowledge of how schools are allocated within Group, you will start to see the trend. At 5th SFG, I knew that school allocations were pushed out to the Battalions around August.

Usually getting the schools for specialty Teams or regular Army schools were easy to secure. However, the specialty schools (SFARTAETC, SOTAC, SFSC, SFISC, ASOT) were slim for the Company.

As a Team Sergeant, I knew what specialty schools I needed to meet 350-1 requirements, and I projected or anticipated my losses due to ETS, PCS, retirement, or the yearly SWCS Levy. This allowed me to identify the schools I wanted to request for my ODA and submitted to the Company SGM, ALONG with my school quota request.

This was submitted to the Company SGM usually around May or June. In all three years as a Team Sgt, I was able to secure the 1x specialty school (SFARTATEC or SFSC) allocated to the Company each year, and further advancing my ODA's capabilities. This, again, was because we were first to submit and planned far in advance.

> Below is the school request (wish list) that I submitted to the Company SGM, along with the School Quota Request (figure 3-2, SFOD-A school wish list). Sending the Quota request was not necessary, because the school was not secured yet, but again, it pays off to be first, as it did with my ODA.

MSG Longoria's method is sound, be sure to follow up with the SGM as often as needed to make sure he didn't "lose" your wish list.

3-5.4 However you request and track schools for your team, there will be courses that fall out of the sky. These schools never show up when it is convenient to make a decision. You will be on leave, halfway across the country when your SGM calls. "Give me a name for _____ school," most likely a super rare slot, or something you absolutely need for your upcoming mission. He will give you the ridiculous time hack of 10 minutes to cough up a name otherwise the slot goes to a team down the hall.

Always have a copy of the Longoria product below, or a hand jammed OML with you at all times. Always remember when you commit a guy to a school, look ahead on your LRTC. Ask yourself, which training events will this guy miss? Is he a critical MOS or a lead instructor? Do I need to request someone to fill his shoes? Does his absence have potential effects on a DTMS score card?

Just as important, if not more so, how does this school affect his life? Did he have leave, a family vacation, an anniversary, etc. lined up? If so, you may need to consider sending the next guy on the OML. This is directly tied to ***Know your men and look out for their well-being.***

SFOD-A 5135 FY20 School Wish List						
Company	Course Name	QTY	School Code (If Known)	Dates	Quarter	Notes
Mission Critical						
C Co	SOTAAC	1		6 Jan - 7 Feb 2020	2	
ODA 5135						
	Special Forces Sniper Course	1	331(2E-F67/011-ASIW3)	6 Oct - 6 Dec 19	1	need for 350-1 requirement
	Dive Medical Technician	1		1-20 Dec 19	1	
Essential						
	Combat Diver Qualification Course	3	331(2E-SI-ASI4V/011-ASIW7)	11 Oct - 22 Nov 19 (x2)	1	
				16 Jan - 28 Feb 20 (x1)	2	
	ASOT Level II (Achilles Dagger)	1		8 Oct - 16 Nov 19	1	
	Combat Dive Supervisor	3	331(2E-F65/011-ASIS6)	~1-20 Dec 19 (x2)	1	Held concurrently with DMT course
	"NSW Course or Key West"			~Jan 20 (x1)	2	
Team Priority						
	Ranger	4	071	15 Sep - 15 Nov 19	1	
If Available						
	Military Free Fall	2	331(2E-SI4X/ASI4X/011-ASI V8		1/2	Will take any available slots
Ft. Campbell Schools						
	Pathfinder	3		~9-25 Oct 19	1	Dates not published for FY20, these are historical dates for FY19
	Air Assault	4	425(SI/ASI2B)		1/2/3	As available

Figure 3-2. SFOD-A School Wish List

3-5.5 **MOS and individual skills training**. Block off time for MOS individual training. You might have to go old school and dig through some Soldier Training Publications (STPs)[1]. Sit down with other Team Sergeants, SGM, and the Battalion schools NCO. They should know of or have a list of schools you may have never heard of. If you still cannot find what you are looking for, reach out to the SWCS MOS committees. See if they know of any MOS specific schools that would benefit your team. Chances are you know a guy in the committee you want to reach out to, or a buddy knows of someone.

3-5.6 Your local Infantry Brigade should conduct Expert Infantry Badge (EIB) testing every year. This may take some coordination through your Group S3 office, but not the worst way to get your guys some reps in basic Infantry tasks. Your medic can do something similar with the Expert Field Medical Badge (EFMB).

If those are not an option, Ranger Stakes are easy to find on the Google. The documents provide Action, Conditions, and Standards. You just need to provide equipment and a place to train.

3-5.7 Snipers have annual requirements. Include time on the calendar, request ranges, and secure ammunition. If you are not a Sniper, have your qualified guys/and or 18Bs

[1] See Chapter 6 for full STP manual numbers.

do the work of digging into the manual for the information.

3-5.8 JTACs and 18Ds are time vampires, but they must maintain certification. Look in a current 1st SFC(A) 350-1 for required certifications. Know when their credentials expire, and request slots early. Once a training slot is secured for 18Ds, look at your calendar to see if they will miss training events. You might have to request a medic to conduct Team training.

Additional thoughts on individual training.

3-5.9 You will have guys who want to take the long walk. If he tells you early enough, this is the block of time I would encourage him to shoot for.

3-5.10 **DTMS**. I am a few years removed from team life, more importantly, DTMS protocols. Get answers to the following: What is mandatory to upload? Who is responsible for uploading all the documents? Do you and the Captain need permission to upload? Get familiar with your Group's policies on what needs to be uploaded such as: individual tasks, ACFT, weapons scores, etc.

Phase II, Collective training phase: Shoot, Move, Communicate, Medicate, and Thrive as a Team.

3-6 MOS cross training. In my opinion this training should come first when you reach this phase. Muscle memory for critical tasks such as: operate assigned radio, individual and truck mounted weapon systems, conduct buddy aid, etc., all while sleep deprived is built here. Do not do this in the Team room. The B-Team will ruin your day every time. If this training is done right, each MOS will produce products for their classes, which can serve as a working lesson plan within the overall POI on JCETs.

CSM Dorsh's input on MOS Cross training:

> A simple and efficient plan to get these knocked out is to go out and spend a couple nights out at the range or training facility – away from the TM Room. Go reverse cycle, put your team into a "deployed' setting and crank out the cross training. This is a great opportunity to focus on the team and collectively come together and bond. This method doesn't have to be fancy but can be extremely efficient in training the basics. Once you get the basics down, make sure your detachment can do it at night, under nods, under stress, etc. Add complexity to the scenarios (leader down, MOS down, etc.).

3-6.1 **18Bs**- Every range they run is checking the block, make sure they produce a quality document they can use in the future (future concepts, JCETs, etc.). When you review the training plan ensure he followed your right/left/and end state guidance.

As the 18Z you have every right to jump in and run ranges, you should have the most trigger time on the Team. Personally, I let my 18Bs run the ranges, it was their time to practice, and improve upon the skills necessary to mold younger guys and teach old dogs' new tricks. Unless there is a major safety issue, I never stopped the range and took over. Generally, I pulled the 18B aside and asked if I could interject. Or I would provide some feedback on teaching methods (at a natural pause or during lunch). If you did your job before range day, you should have known his plan and improved it prior.

The flat range is not the only live fire range an ODA can use. Don't get me wrong, it is a fantastic tool that can be utilized in many different ways. The 18Bs need to schedule and train the team on every weapon system expected to be used overseas.

- Belt fed weapons (mounted and dismounted)
- Mortars
- Shoulder fire
- Foreign
- Recoilless

Every US weapon system has a qualification table for day and night. Your 18Bs should know which regulations to reference... and do not let them forget about the CBRN tables.

Do not discount the old-school pop-up qualification range. It will demonstrate the true zero for their optic selection and/or setup on their primary weapon. Also, a great place to knock out CBRN qualification check lists.

Every installation should have ranges designed for individual, buddy Team, squad movement, and movement to contact ranges. These are invaluable for testing your SOPs and building trust across the detachment. Yes, you can use a flat range for this type of training. I would agree that a flat range is a good place to start. Everyone can see each other; no obstacles or terrain features to block your view as a leader directing traffic. As the team progresses you can add barriers and vehicles for cover, place targets to force weak handed shooting on the move, conduct cover and move... what you are missing is terrain. Imperfect rolling terrain, with vegetation that will trip a guy up under NVGs.

Fort Carson Range control executed legit range walks, so you knew what every range could support. They are there for you, use them.

18Bs are the Detachment's tacticians. They should be able to put together a class on base defense for the team. Almost everywhere we go, a base defense plan can, and should be put together.

3-6.2 **18Cs**- They know how to do more than breaching charges and property books. Make them cross train a plethora of initiation systems, establishing ring or line mains, a variety of charge construction based on target requirements. Knowing the Commander's Annual Training Guidance, the Battalion's LRTC, and your experience can help drive what demolitions training your ODA needs to conduct.

This might require some intense planning but, have them build forms and pour concrete at the demo range. Maybe haul in and set telephone poles for timber charges. Acquire different types of metal, bring in junk cars. Group Engineers should be able to help, or Range control might have resources for this, ask questions.

Also make sure your 18Cs know how to set up, run, and conduct basic maintenance on generators.

3-6.3 **18Ds**-Self-aid, buddy aid, 18D... this briefs well as the medical PACE plan but if someone does not know how to do the basics with their dying buddy's IFAK, then that plan is crap. LTT is the best option for medical training.

> (EL) Ft. Campbell has training sites called RASCON (school of advanced medicine). This training site has all sorts of automations and scenarios that you can create (smells, day/night, noise, automated casualties that are controlled with a device-respirations, bleeding...) This is a great site to train and validate your TCCC. Conveniently, this training site was located directly across the road from a range and urban site. As we conducted training and simulated a casualty, we would work our way to RASCON and continue with treatment using the advanced casualty mannequins.

Fort Carson had a similar facility, but not conveniently located near an urban range. Although, make sure to ask about the mount site just over the hill by the Group compound, I believe it was range 60. The range guys always were selling their ability to recreate certain aspects of the battlefield.

3-6.4 **18Es**- Due to technology growing exponentially each year, and not wanting to screw the pooch on OPSEC, I am going to keep this section light. For specific individual and collective tasks for the 18E dig into 350-1, Chapter 6, para 6-9 to ensure your team is ready to meet the communication challenges in your AOR.

Even with new hotness coming down every year, don't forget the basics. Everyone on the Team should know how to operate their individual radio. This is a must, the 18E does not have time to set up everyone's radio and correct basic issues such as a dropped fill. Additionally, all team guys need to know how to set up, run, and conduct basic troubleshooting on truck and TOC radios systems. All members should have a working knowledge with the other platforms the team utilizes.

This is the time (collective training) to use Army radios for every training event you can. Does a flat range require radios? Not really, but you drive there. Require the Team to set up truck radios, work through convoy radio SOPs, "Right turn open, right turn closed, checkpoint 3 open... CP 3 closed." Build this muscle memory before the FMP phase.

3-6.5 **18Fs**- Again OPSEC limits what I can cover here, but your 18F can run the team through things like Sensitive Site Exploitation (SSE) collection methods. He can cover how he incorporates OAKOC into planning, and how to conduct a PMESII/ASCOPE cross walk during IPB. PMESII/ASCOPE is a simple concept but can be confusing right up front. Have the other MOSs dig into the areas they will be responsible for and learn how to do their part during IPB.[2]

Another big area your 18F needs to cover is Threat Vulnerability Assessments (TVA). Everyone on the team should understand each locations strengths and weaknesses and the plan for each. If you are short on time, at a minimum get your 18Bs and 18Cs looped into the TVA process. They fill special roles in the assessment and developing/implementing the plan to reduce the threat.

During this phase of training the 18F should be done with his country study for your upcoming trip. I am pretty sure they are not called that anymore, but for this book I am bringing it back! He should start briefing the team a current intel picture once a week (if your destination is lively enough, maybe a daily SITREP is needed). Even if the trip appears to be low threat, the country study will help everyone understand all that is required to dominate in the human domain.

---

[2] See Annex I, pages 297-307 for a detailed IPB (AO, AOI, AI, what, so what, now what, OAKOC, VWCPT, etc.) breakdown.

3-6.6 **Enablers** Do your homework on your upcoming mission. Is the norm to have attachments? If so, which MOSs? How have they assisted in mission success? Were they a distraction? You can find answers to the above questions in post mission products. Get with your unit Lessons Learned contractors or dig around the portal, shared drives, and talk to other Team Sergeants.

If they are needed for your upcoming mission, and they do not present any safety concerns for your training events, start to incorporate support MOSs before the Full Mission Profile phase.

When you are preparing for a trip that requires outside support, think of ways to train these Soldiers. By doing so, you can create a good name for your ODA within the support units. For example, as a Mountain ODA when feasible we would run our enablers through the basic mountaineering task list. At the end, if the enabler was able to complete the tasks to standard, we would produce a certificate of completion. The certificate served as promotion points for the SM, and now the ODA had a support Soldier that was familiar with our infill specialty. This extra effort will pay off when you need last minute favors from these MOSs.

Enablers deploy with us all the time. Why not learn who is in the support companies? Train them so when the time comes, you can request by name who you want down range with you.

When our Company was going back and forth to Iraq, I was familiar with the AOB and/or the SOTF having most of the supporting MOSs centralized. They would visit ODAs, or we could push equipment to them to fix.

Later in life, when ODAs were pushing to austere locations in Africa, we needed to bring those MOSs with us. The above paragraphs became the standard in 10th Group. I can say this with a degree of certainty, a majority of the soft skill guys/gals were dang good at their job, and loved doing it. It was rare to receive a dud who needed to be swapped out.

It was imperative to counsel him/her when they arrived to ensure they understood their role. On the other side of that coin, I sat the guys down to tell them that the support kids were not our squires.

I'll close this topic with CSM Dorsh's input:

> Do not forget to build the team with your enablers. Don't treat them like 2nd class citizens, outsiders. Embrace them and make sure they understand the detachment's culture and what values the ODA stands for and what they will not put up with.
>
> Most enablers will work very hard to not let you or the detachment down, give them the chance to carry their own weight and earn their way. It will pay off in dividends down the road... maybe go to

selection one day because of the way they were treated on your detachment.

3-7 Collective training, not a lot to cover here. You and the Captain have already identified all of the collective tasks the team needs to practice before your trip. Lay these tasks out in a logical manner throughout this phase on the LRTC. Each event should build on the skills used, with skills yet to be covered. Lastly build them up regarding longevity, complexity, and difficulty.

Phase III, Full Mission Profile (FMP) phase: Validation events.

3-8 If you have the ability to create your own FMP in order to validate your team, use the DTMS score cards to aid in planning the event. Prior to combat rotations this will probably be planned, coordinated, and resourced by higher. More than likely the event will be conducted at a Combined Training Center (CTC) location:

- Joint Readiness Training Center (JRTC)
- National Training Center (NTC)
- Joint Multinational Readiness Center (JMRC)
- Fort Bliss Operational Readiness Mission Training Complex.

The ODA shouldn't have to directly coordinate for evaluators. You and the Captain need to follow up with Battalion to ensure this task has been done though. The grade sheets go into DTMS, and Commanders will rack and stack ODAs commiserate with their performance during the validation events. ODAs who perform well, will be aligned with the best missions within the AOR.

If our nation is engaged in combat operations, your Battalion and/or Group Commanders will select the best ODAs, based on validation events (and other factors) to meet the Geographic Combatant Commander (GCC) and TSOC requirements.

Depending on your Group, mission, TSOC/GCC, and role on the strategic map you may need to dig into the Certification, Verification and Validation (CV2) pathway. This requirement came front and center, and changed how ODAs are certified to deploy, after the deadly ambush in Niger.

3-8.1 **Validation focus**. FMP reps should be hyper focused on the most critical, difficult, and dangerous tasks for whatever mission you are preparing for. If you are going to the middle of nowhere Africa, and know you will not have helicopter assets, do not waste time planning and coordinating FAST roping onto the X scenarios.

Not only will you be denied by the boss, but you will also lose most if not all of your credibility as an ODA. Your FMP should be designed to closely replicate what you expect to do on the ground. As a Special Forces line Company detachment, your OCONUS mission will include a partner force. Nothing earth shattering here, at a minimum you will train, advise, assist, and with any luck accompany them. Work as hard as you can to get a role player partner force for your FMP.

3-8.2 **It's been a while**. If it has been a hot minute since your last UW training event, or you have never planned a UW FMP, crack open the manuals. The top three should divide and conquer here. There is a lot to absorb, start early, and have more than a few brain-storming sessions. Between the manuals and DTMS score cards, a clear path to your FMP should emerge.

3-8.3 **Back in the Day!** Thinking way back to one of my Pre-mission Train-ups (PMTs), when the leadership got it right. Background: 5th and 10th's SOTFs were high fiving every six to nine months in Iraq prior to, and during the surge.

After both individual and collective skills training phases, the Company leadership put together a realistic FMP. This FMP was not to be run by the Team Sergeants and Captains. The senior NCOs on the teams were expected to step up to the plate. At the time, most of the Company was comprised of guys who had been there a few times. Many were headed back on their third or fourth trip.

As one of those senior E7s, we were responsible for developing a Partner Forces assessment checklist. From the checklist we created an adaptable Program of Instruction (POI). This POI covered individual and collective tasks required to prepare Iraqis for combat patrolling in a COIN environment. The Company

Commander ensured we had the correct doctrine at our disposal (back before Army PUBs was a thing, or before we knew about it) to base our POIs on. The detachment leadership was there to help coach us through our development process and ensured risk mitigation measures were followed. We had to prepare and brief our plans to the Company Commander as if we were the detachment leadership.

This PMT was as close as it gets when it comes to replicating down range. Our Partner force was comprised of some Infantry guys, but most of our role players were low density MOSs from units on Fort Carson. My memory is failing me, I believe we only had three weeks to assess, train, and begin simulated combat operations.

We started on the flat range, with focused efforts on the fundamentals of weapons safety. After weapons safety, we moved on to static shooting drills, individual movement drills, and buddy team movement. Once we left the flat range we had transitioned to blank and paint rounds. Our POI continued all the way through to squad and platoon mounted and dismounted patrolling in an urban environment.

Each ODA was assigned a platoon (30 or so Soldiers) to advise, assist, and then accompany. The goal was for the teams to train their Partner Force separate from the other ODAs, then come together for larger scale operations with multiple ODAs.

For the large-scale operations, the ODAs and their partner forces were based out of Camp Red Devil on Fort Carson (Red Devil is a small fenced-in compound near a drop zone with a flight landing strip, and a fairly large urban training area). At the camp we had three ODAs and Partner Forces sharing space.

Between the three Groups, we shared everything from providing security, to cleaning details. Acting Team Sergeants de-conflicted vehicle marshaling, bunk space, common areas, and the set-up of a Tactical Operation Center.

Once the joint combat outpost was occupied, we began planning operations. Every mission required us (the ODA) to take the back seat while putting the partner force in the lead. We advised, assisted, accompanied, and conducted post mission activities, just as we would overseas.

I cannot remember the number of missions we conducted overall, or how many days we stayed there. I do remember two raids on Fort Carson proper, and the final hit at the Pueblo Chemical Depot. The operations on Fort Carson were ground assaults, with Apache gunship support. For the Chemical Depot raid, we jumped on CH-47s, and I believe we had the Apache's again that night.

Every mission highlighted what we failed to cover in training or proved that a few weeks is not enough time to completely hand over the reins. It showed us that a balance was necessary to put the Partner face on the ops, while still maintaining control for overall safety and success on the battlefield.[3]

3-8.4 **Mission Products**. Write and use WARNOs, OPORDs, FRAGOs, or CONOP templates to execute FMPs. Not all CONOP templates are created equal, review examples from different Teams, see which ones cover the basics from TC 3-21.76, and contingency planning. If you can get your hands on it, use the CONOP template your upcoming mission's GCC and/or TSOC expects you to use.

After enough reps of the top three writing and issuing mission orders, have senior guys execute. By doctrine Staff Sergeants, and Sergeant First Classes should be able to advise, assist, and accompany partner force Company level leadership with everything from planning, through to execution of combat operations.

One of the most important benefits is your 18F gets reps building and briefing the Situation[4]. Provides him the opportunity to build out the Area of Operation, Area of Interest, and Area of Influence. He should answer the <u>What</u>, <u>So What</u>, and <u>Now What</u> for:

- Weather/VWCPT (Visibility, Windspeed/ direction, Cloud cover, Precipitation, and Temperature)
- OAKOC

---

[3] We did not execute full CTC rotations for unit validation prior to deployment at this time. I believe it had to do with the frequency of the deployment cycle.

[4] See Annex I, pages 297-307 for a detailed IPB (AO, AOI, AI, what, so what, now what, OAKOC, VWCPT, etc.) breakdown.

- MLCOA
- MDCOA.

Some guys can get sucked too far into the "big picture." They will work way too hard to build a robust Situation paragraph in an OPORD, or within IPB. Monitor closely, ensure they are answering the what, so what, and now what during each, and how those factors can, and will directly affect the ODA.

3-8.5 **Last minute ankle biters**. Whether you knock it out before or after your FMP phase, ensure you have a plan to be mission complete on admin stuff before you deploy. This includes language scores. Language testing is one of the admin things you have some control over. What I mean is you have a little room to maneuver everyone's test date, if you know when they tested last, you can have them test again if it has been six months or more.[5]

3-8.6 Last thought on the LRTC, never let the 180A and/or the 18A build out the LRTC without you. If any changes had to be made (operational changes/updates, etc.), while you were out, get your eyes on it before they submit. You are the sanity check for the detachment.

3-8.7 **The Longoria Training Development Model.**

> (EL) Back-story: In my three years as a Team Sergeant, our ODA was always ahead of the curve

[5] Review the policy on testing frequency, the SM may be required to complete a set number of training hours prior to scheduling a new test.

with all training concepts, and training support requests. The above was submitted so far in advance (approximately eight to ten months) that often the BN would forget to fill support requests. As any good NCO knows, follow-up conversations or phone calls with Battalion S3 was essential.[6]

Additionally, in my three years, my ODA only filled three taskers during the Battalion's eighteen month (collective) Red Cycle Tasking period. Yes, this was an unimaginable amount of time to devote to a Red Cycle tasking period between two deployments, but the taskings we had to fill were simple. Two were running the Group Maritime Assessment Course, which we were providing instructors anyway because we had teammates in the class. The third tasking was one we volunteered for as it would showcase the ODA. This was a 4-hour range for the city of Clarksville Mayor and entourage. Of course, this was high visibility with all Group/Battalion leadership present.

I was able to achieve this due to instituting a Team PLANEX. Essentially, after digging into the Annual Training Guidance, deployment playbook, mission letters, and AOR developments, we developed our training

[6] 10th GRP note: I had to route most of my traffic through our B-team/SGM. Calling Battalion directly was frowned upon.

calendar as an ODA. The Team leadership would first assess the ODA, list out the strengths and weaknesses. Then, the leadership would choose which direction we wanted to take the ODA. Following this, the Team Leader provided his intent. I would produce a blank calendar twelve to eighteen months out (no 180A my entire time as a Team Sergeant), all the way up to the next deployment or playbook change.

I filled in the planned and tentative USASOC training holidays, leave dates (summer and winter), tentative deployment dates, and PMT's. I also provided a list of ATRRS dates for the schools (as discussed earlier in this chapter, Individual skill training phase) we needed to attend to meet 350-1 requirements along with then the "nice-to-have" school dates.

The remainder of the ODA was split into two planning cells (minus the 18F as he served as the red hat during development), given training guidance from the Team Leader and my broad stroke training objectives. The planning cells would then conduct Course of Action Development (COA DEV), starting with individual training all the way through to FMP. No restrictions regarding venue, or concept ideas were placed on the planning cells.

Upon completion, each cell delivered their plan to myself, and the Team Leader, with the 18F red hatting the training plan. Usually at the end of the day, as with all COA DEV, we would take the best parts of each plan, insert the PME school dates (by individual or basically a Troop to Task on the calendar), and generate the overall training calendar. I would consolidate the plans onto a general calendar and provide to the team leader for his additional guidance, if any.

Following a quick sanity check to ensure we were not too eager or reaching burnout, this was posted and became our training calendar for the ODA. I assigned Primary (PI)/Alternate (AI) training instructors to develop training concepts with No Later Than (NLT) times for review and submission. Finally, I gave my left, and right limits, expectations, and training objectives for the PI/AIs to develop their POI's. Of course, I supervised all aspects of this and served as the Contingency for all training concepts in the event something fell through. I usually only stepped in when the Team ran into major roadblocks and needed some rank weight to assist. Otherwise, the SSGs and SFCs on the ODA developed their training concepts, conducted all of the coordination's with training vendors, or with support personnel.

This method allowed a few positive aspects for the ODA.

- Developed the young guys, they learned how to plan, coordinate, resource, and develop training concepts.
- You and the Captain are not jumping through hoops trying to secure challenging, realistic training for the Team (or tagging along on some other Team's training concept because you failed to plan).
- Allows time for the Team Sergeant, to ensure all backside support requirements are met or enabler support needed to enhance your training.
- Provides the ODA, and their families a level of certainty with a planned-out calendar.
- MOST IMPORTANTLY, this method gets "buy-in" from EVERY MEMBER as they all developed the training calendar together.

Team assessments and benchmarks were then monitored as mentioned in the Marlow Product (Figure 3-1).

Below is the Training Tracker that I developed (Figures 3-3a, 3-3b). The tracker ensured that all support requests, and enablers for each training concept was not missed. This was a living document that was posted in the team room. The assigned PI and/or AI would update the tracker as requests, and requirements were filled, they changed the color of the box, and added a date or status.

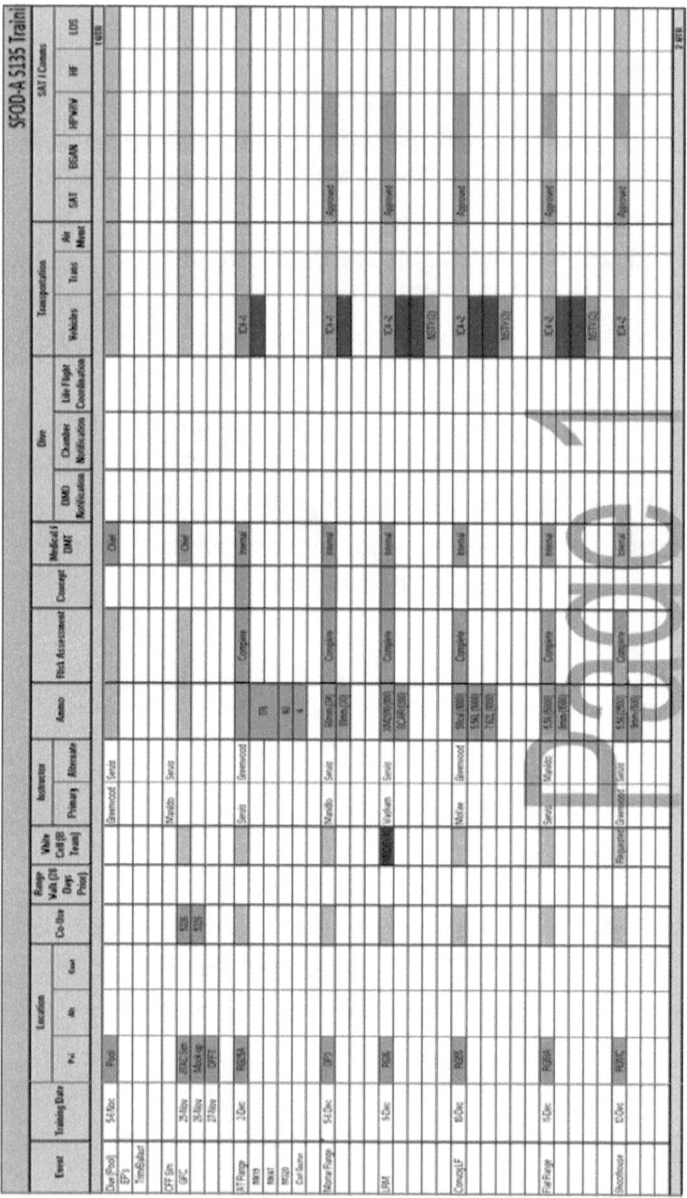

SFOD-A 5135 Traini

Event	Training Date	Location			Co-Use	Range Valb (28 Days Prior)	White Cell (B Team)	Instructor		Ammo	Risk Assessment	Concept	Medical / DMT	Dive			Transportation			SAT / Comms				
		Pri	Alt	Cont				Primary	Alternate					DMO Notification	Chamber Notification	Life Flight Coordination	Vehicles	Trans	Air Movt	SAT	BGAN	HPW/IV	HF	LOS
																								1 QTR
Dive (Pool)	5-6 Nov	Pool						Greenwood	Servis				Chief											
EP's																								
Trim/Ballast																								
CFF Sim								Manildo	Servis															
GFC	25-Nov	JTAC Sim			5326								Chief											
	26-Nov	Mock-up			5326																			
	27-Nov	OFFT																						
AT Range	2-Dec	RG25A						Servis	Greenwood		Complete		Internal				ICA-1							
MK19										576														
MK47																								
M320										40														
Carl Gustav										4														
Mortar Range	5-6 Dec	OP3						Manildo	Servis	60mm (24)	Complete		Internal				ICA-1			Approved				
										81mm (30)														
LRM	9-Dec	RG6					[illegible]	Vasham	Servis	[illegible]	Complete		Internal				ICA-2			Approved				
										SCAR (100)														
																	NSTV (2)							
Convoy LF	10-Dec	RG25						McKee	Greenwood	50cal (1000)	Complete		Internal				ICA-2			Approved				
										5.56L (1000)														
										7.62L (1000)														
																	NSTV (2)							
Flat Range	11-Dec	RG30A						Servis	Manildo	5.56 (5000)	Complete		Internal				ICA-2			Approved				
										9mm (1500)														
																	NSTV (2)							
Shoothouse	12-Dec	RG30C					Requested	Greenwood	Servis	5.56 (2500)	Complete		Internal				ICA-2			Approved				
										9mm (500)														
																								2 QTR

Figure 3:3a Training tracker

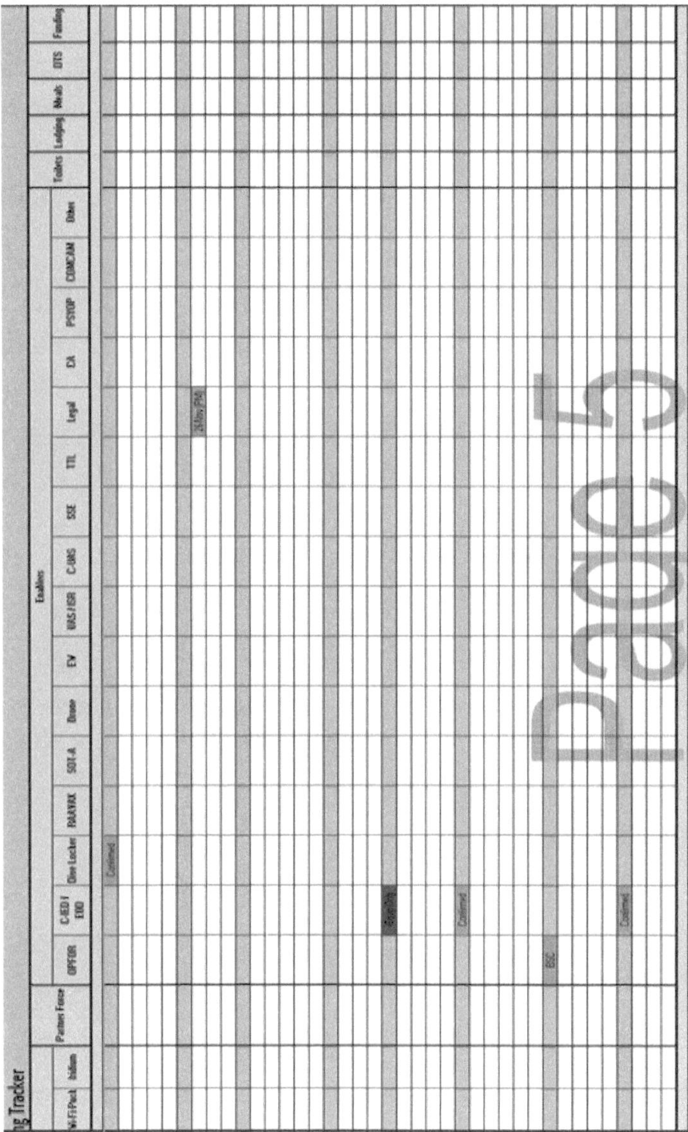

Figure 3:3b Training tracker

3-9 Hour by Hour Tips (6-9 Week Training calendar). As mentioned in Chapter 1, in the Authorities and responsibilities paragraph, your control measure for the ODA is the 6–9-week training calendar. A well prepared training calendar is also your initial defense against taskers.

The easy button for Battalion trying to fill taskers is selecting the ODA who has nothing scheduled. If you are creative, you will never have un-planned time. Hold on a second! Earlier I recommended to include “white space” into your LRTC and hour by hour calendars. True, but if you are in the collective training phase for example, you should include cross training in the afternoons during the admin week.

For this tactic to work, you must be able to articulate the criticality of this cross training to your SGM. He needs to understand how the cross training is directly tied to the next training event, or the upcoming FMP/validation event.

If you are consistently ahead with your hour-by-hour calendars, your white space is less tempting for Battalion to fill with taskings.

Looking back a few pages in this chapter, under **The Longoria Training Development Model,** he does a fantastic job explaining his experience with this topic. His ODA was able to demonstrate that if you put in the effort, you can almost own the entire year. In his case, they only executed taskings that truly benefited the detachment.

3-9.1 Have a white board near-by wherever you hang the hour-by-hour. These calendars only have so much space available, it is nice to have a place to provide additional instructions. For example, Task- "range prep". Do we have all the required range control documents ready? What status is the range, reserved or pending? How much ammo are we drawing, and do we have those documents ready? Which truck does the ammo go on? Who is conducting the ammunition draw?

3-9.2 Do not waste your time, ask for a calendar template approved by the SGM. You will spend many hours creating and updating these calendars. If you were unfortunate enough to be issued human emotions, then you will be cut deep when you have to re-do all of that work... Because we all know the new format will not accept cut and paste.

3-9.3 At the end of the day you are required to build these calendars, make them work for you. Include enough information on them so your team can continue forward with the work if you are pulled away.

3-9.4 **Physical Fitness**. Common sense says our job requires us to maintain a high level of readiness in this area. That being said, the team's high op tempo, Company training meetings, random meetings because someone forgot to put something out at the last meeting (all things that could have been an email), family requirements, late night concept development parties can slide physical fitness onto the back burner.

I missed the mark with physical training. I did not select and utilize specific PT programs directly geared to prepare the team for our ODA's assigned OCONUS mission. Instead I bounced around between team sports, the Manitou Incline, trail running, and random gym routines.

THOR3 has been around long enough to know what we do, and how to build PT programs to prepare us for them. Take the time to sit down with the training coaches. If you can summarize the physical requirements for your upcoming trip, they can build a program for your team.

12-18 months is a long time to train for a specific fitness goal. Your LRTC should already be broken down into three training phases, use these phases to build a common-sense fitness program. Use a progressive program that builds a solid foundation during the Individual Skills phase. As you push into collective training and ultimately FMP phases your Op-tempo should increase, and your PT program should become more and more focused on your specific fitness goal. Lastly, be prepared to get creative when scheduling PT into the training week.

Most of your guys will continue to crush PT without your input, but a few will not. You are responsible for the performance of the Team. I strongly suggest you make the time to conduct Team PT. Do this once a week at a minimum, just to keep you and your guys honest.

Otherwise, if you send a dude to a DA Form 1059 producing school, and he fails the ACFT, you are on the hook for that failure. Now the Battalion and Group CSMs know who you are in a bad way. If one of your guys fails to meet fitness standards at one of these schools, one can argue that it is just as detrimental to a career as a DUI, or a felony.

If you are looking for a fantastic breakdown on physical fitness and the overall importance for individual and team functionality pick up a copy of *Always Endeavor*. In Chapter 3 "A Lifestyle of Wellness" page 47-76 has a lot of good information on this topic. Colin Greata does a great job tying Physical fitness, Nutrition, Cognitive development, Rest/Recovery, and Emotional wellness together into his Wellness Pillars.

3-9.5 **Monday.** It is not a bad idea to keep Mondays free during most training events, first example being range weeks. In my experience, scheduling range time (minus SFAUCs, or multi-week events) on a Monday is a bad practice. Nothing ever goes right on Monday morning. Issues at Range control, the ASP/AHA, Motor pool, etc., always devour training time.

3-9.6 **Potential time saver**. If you are executing a reoccurring training event, like Rifle qualification,

Language training, etc., the hard work is already done. Find that hour by hour in your digital library, change dates, the enclosure info, and whatever else is required, then move onto the next product.

3-10 Training Concept tips. Training Concepts are a contract between you and the Battalion Commander. Once the packet (full 350-1 concept, one slider, etc.) is signed, the boss expects your ODA to execute it to the best of your ability. On the other hand, you expect it to be fully supported. Keep your concept resource trackers straight, and do not assume that Battalion cares about your training as much as you do.

3-10.1 **Research**. Never re-create the wheel, spend time on the Group's local drive, portal, GLOCS (10th Group concept tracker at one point), etc.

Search for training venues that check as many DTMS boxes as possible. Ridge Runner, HAVE ACE East and West, USMC Mountain Warfare Center in Bridgeport, CA, etc.

Dig into other ODA's training concepts for inspiration. Look at your Group's run Achilles Dagger training events, here you can check some blocks in UW training.

3-10.2 **Cite Regulations supporting your concept.** It may seem trivial, but this gets overlooked a lot. There are many times you will submit training concepts/one sliders where all you do is change dates and names. Be careful here, our doctrine changes often. Be sure to confirm you are using the most current reference(s).

When the doctrine changes, be sure to read, at a minimum, the "Summary of changes" page if you find an updated regulation before you cite it as a reference for your concept.

3-10.3 **Training locations**. Look outside your installation for training opportunities/locations. Example, Guernsey, WY has a National Guard camp with a full complement of ranges (small arms, belt fed, mounted, recoilless, and demo). The training area also has a drop zone, an airstrip that supports RAPIDs, C-IED lanes, Conex cities, etc. This could work for an Emergency Deployment Readiness Exercise (EDRE) training event.

Fly away training events are the best way to test your systems and checklists. Once you are on the ground, there is no running back to the Team room for a forgotten item. Another bonus with a fly away is the B-TEAM can call all they want, but you cannot break away from training to attend a meeting.

3-10.4 **This is the way**. Training concepts for the collective training and FMP phases, especially the FMP phase, require a Partner force. Fight like hell to get one; infantry, soft skill units, GSB Soldiers, it doesn't matter who. Support MOSs are in my humble opinion the better option. I covered this previously, but worth saying again. They will truly test your POI, but more importantly your guys. They provide a realistic skill and fitness comparison to most of the partner forces out there. Infantry units live, eat and breathe small unit tactics. With support MOSs you must focus on the basics and keep it simple.

GBs create combat power! Nothing against the CRF mission, but do not fall into the unilateral mindset. Outside of a few things I shouldn't go into, ODAs are not designed to be a unilateral force. Also, remember we are the Regiment's greatest recruiting asset. This is the time to shine and attract the best from the big Army.

3-10.5 **New equipment**. Make friends with your Group S8. He receives new stuff to test all year long. If you have specific items you are looking for, reach out to him. Tip, if you can tie your equipment request to the Annual Training Guidance/Mission letters you might have a better chance of securing what you want.

3-10.6 **Quality control**. Develop and use a SOP for all products leaving your team room. Specific to Training Concepts though, develop your SOP around what Battalion and Group want to see. If you have taken the time to dig into our 350-1, you will find there is the baseline standard for training concepts. Every group will add to/take away from this, know what those changes are.

It takes time to create concepts, do what you can to reduce the back-and-forth game between you and the approving authority. Every product that leaves the team room, speaks to your team's attention to detail. Once more, the boss does not get a lot of face time with your ODA. Your team's worthiness will be gleaned from your products submitted to higher.

Something to consider, a method to save the Captain's, 180's and your time is to have your guys do their best work before handing it in to you. It is one thing to clean up some knuckle dragger sentence structure, but to have to do a complete re-write hinders your ability to manage everything else on the team. Anything that brings your attention down to a single task limits your ability to manage the whole team.

3-10.7 **Don't miss out on the same training opportunities I did.** Thinking back to my first six months or so of executing training concepts, I remember way more chaos than smooth sailing when getting out the door. Not a jab at my guys, they were good, and very rarely did they not have what they needed to conduct training.

Where we were not synced was at my level. Had I started my Team Sergeant time using basic methods, and tools (Ranger handbook, checklists, info boards, etc.) to plan, prepare, and launch training events, then this paragraph wouldn't exist. It took some time before I even executed truck side convoy briefs and implemented the circle of trust (PCIs/PCCs).

As I wrote the above paragraph, I asked myself why didn't I implement the basic methods, or use those tools? The best I could come up with is that I didn't want to insult my guys with those. Or that I didn't want to look weak as if I didn't know my job, or how the team functioned.

In Kyle Lamb's book, *Leadership in the Shadows*,[7] he tells his story of growing up in the 82nd. As a young paratrooper he explains how the Jumpmasters had the pre-jump briefing memorized. This became his view of what a professional looked like. Once he made it to Group, during his first jump the JM was reading the same pre-jump brief from the book and expressed that he was not impressed. After a while he came to the conclusion that leaders in SOF have a lot going on. One day you are jumping static line, the next could be Freefall, or rolling into a SFAUC. Due to our large breadth of skill sets, it is imperative that leaders have a tool (checklists) to ensure mission success and safety during all events.

To avoid my above mistakes, take the following paragraphs to heart. It may seem corny at first, but if you - the boss - take it seriously, then the boys will adopt your mindset. The sooner that happens, then everything the team plans, prepares for, and executes will be taken seriously.

Conduct a mission brief prior to every team training event and incorporate checklists whenever possible. Whatever your planning products are, use them to brief the men before heading out. When your training is a multi-day, or multi-week event close each day with a FRAGO, so the

---

[7] Leadership in the Shadows, Ch 36 Wish list or Checklist

team can prepare equipment for the next day's training event(s). Develop good habits here.

Start with the small training events to develop and test the ODA's planning products and pre-mission activities. When you reach the big training events you should be in the refining mode. If you do not know where to start, dust off that Ranger handbook, open to Chapter 2 and run down the list. Even now, I have mine open (Ranger HB dated April 2017).

Does every trip to the range deserve a full OPORD, no. Do you always have time to develop one, no. At a minimum you should execute the truck side convoy brief. Cover the basics to get the ODA there: route plan, training to be conducted, communication plan, and the medical plan from the information in your training concept. Additionally, you should review MOS/training checklists (PCCs), and conduct a circle of trust (PCIs).

3-10.8 When you do have the time to prepare a full OPORD/mission brief, do it right. Below I went a little overboard providing inspiration to get you started.

You: roll call, references (Maps, overlays, ATAK, Blue Force Tracker, GRGs, Ranger handbook, USASOC regulation x/y/z, Range control regulation, TMs, etc.), time zone used throughout (L does not mean local time), a quick Task organization (make your #2 do this after they have seen you do it), orient, box, trace, and familiarize technique for your training area/target on a map (only the areas the team is moving through).

Situation. 18F gives a quick description of the Area of Operation, Area of Interest and if applicable, areas outside the AO that can Influence your training event or operation. Area of Operation breakdown: Terrain in OAKOC format, and how it will affect both friendly and enemy, be sure he uses a map to cover this.

Weather and its impact on the operation. Enemy forces? If it makes sense with your training event have your 18F tie the Enemy situation to your Team's upcoming OCONUS destination. If not, consider using MPs, Range Control, other units who will rat you out for SF training, or is there potential for a Command team visit?

He should know the basics of composition, disposition and strength for the "enemy forces"; recent activities (where known or suspected speed traps lay), describe known or suspected locations and capabilities. And of course, MLCOA/MDCOA. Again, all briefed from the map.

You and your senior guys, that have been stationed at your Group for a few years, should be able to help develop the pattern of life and TTPs for the assigned "enemy".

Friendly forces, Higher headquarters mission, intent, and concept (one and two levels up). I would use this as an opportunity to connect the Commander's Annual

Training Guidance into the OPORD (make the Captain do this).

Mission of adjacent units, large units shut down roads for PT on most installations. What are the peak traffic times and locations on the installation? Will this affect your route planning? Are you sharing the range? Are the ranges next to you being used, if so by whom? Are you set up for a Co-use of the range with another ODA, or another unit all together?

Attachments and detachments. If more detail is needed here after your task organization has been covered, be sure to notify your team who your attachments are. Examples: Dog team, EOD, SOT-A, EW, CA/PSYOPs, etc.

Mission. The Captain can and should write a mission statement.

Execution. The Captain also needs to write his intent and develop the concept of the operation (think phases of the operation). Nothing fancy here, but this should tie into the ODA's big picture. For example: We are conducting training event X, in order to build to this larger collective task, or towards the end goal of X deployment. Even the basic flat range can be tied into the strategic picture if you try hard enough.

Scheme of movement and maneuver, this is your bread and butter, you and the 18Bs live for this paragraph. Without re-typing table 2-9 - "Example of a squad

OPORD" in the Ranger Handbook - here are some things I would like to highlight.

This paragraph is the operation told in story form. Brief the mission as you have planned it, start to finish. Talk about what the team is doing during each phase down to the nitty gritty detail. Ensuring each member knows their primary, secondary, and possibly tertiary responsibilities. More importantly they know what the special teams are (primary and alternate), and who is on those special teams. Who has what special equipment, and who is responsible for said special equipment if that brother goes down.

Contingencies, I have seen and briefed these two different ways. One, contingencies are briefed at the end of the scheme of movement and maneuver paragraph. Two, they are briefed along the way. Also, I like to keep location of medics/medic leadership/casualty or dead evacuation methods (vehicle or foot movement to closest HLZ/AXP), Casualty Collection Point(s), and TRPs as part of this paragraph. Some like to brief it separate in the sustainment paragraph/later in Execution. Having your pre-planned HLZs, AXPs, briefed in conjunction with TRPs paints a better picture for you and your guys.

The remaining items in this paragraph can be briefed in order after the "story" has been told. Totally up to you. For example, I preferred to hold the timeline until the very end of the brief. Here I could highlight critical/key time hacks, and I made sure it was posted in a common area.

FOOM boards. Feel free to knock these out in this paragraph or hold them towards the end before rehearsals. I liked to cover the vehicle task organization upfront, so the team members can visualize where they are as the mission is briefed.

Sustainment. This should be an easy paragraph for a basic range day. Obviously the 18 Chuck owns this, but will need information from the 18Bs, Ds, and possibly the Z. Why the Z, because you might know there is an arms room inspection next Monday, and yes you need to stay late on a Friday to clean guns. All in an effort to ensure your team isn't the dirty birds of the Battalion.

Nothing mind blowing here, this paragraph is where the Maintenance plan, and DX times/locations are briefed. Spell it out by phase of the operation. This plan should be in accordance with (IAW) your team's SOP, by event, or prioritized by leadership. If your SOP has not been established yet, use this time to develop one.

Every good 18B will bring a magical box of spare gun stuff (lights, optics, slings, etc.), but does your lone 18B, who was a junior just last week, know the proper form(s) for turn-in/requesting specific maintenance or new items through the arms room?

What is the timeline for maintenance post training event? What are the maintenance priorities after training? Is it just a rust and dust immediately after the training event, then a deep clean the next day after PT? Do the trucks just need fuel, or do they need to be returned to the motor pool? Does your Battalion motor pool require a full PMCS when you turn trucks in?

Classes of Supply, does your 18C know all of them? Does he know how to request them through appropriate channels? Better question, does the 18B know which class of supply his targets fall under? Is the unit, or Range Control responsible for providing targets? But at this point you are giving a brief to get out the door. Too late for requests, but here Class I, III, V, VII, VIII, and IX are briefed for pre and post mission activities.

Build good habits with the small training events. Every piece of equipment you take out should be cleaned or at least inspected for damage upon return. If damaged equipment is found, make your guy's DX it immediately.

You need to set the standard here, always lead by example, and follow up with spot checks. If you do it the guys will naturally adopt this method. With any hope they will carry this, and many good habits forward into their Team Sergeant time. Nothing worse than saying "I'll get to it later," then the SGM conducts a surprise Company level competition requiring full kit. Loose/busted body armor is never good during a stress shoot.

Field services, try to make this fun. Can you get the gut truck to come by? Have pizza delivered? Coordinate with the B-TEAM to bring out hot chow from the chow hall, or everyone's favorite, MREs. Careful with food from the

Army, depending on your Commander's mentality you may have separate rations taken out of your paycheck.

Army Health System support. If you chose to brief the MEDEVAC plan by phase of the operation during the execution paragraph, then your 18D only has a few more things to cover like:

- Follow on care facilities listed by trauma rating
- Methods of marking patients
- How wounded/injured will be treated (Self-aid/buddy aid/18D)
- Preventative medicine
    - Note on preventative medicine, the 18D should not be the full-time supplier for sunscreen, chap stick, bug juice, etc. Add that to your team's personal range bag and Ruck mandatory items list.
    - Yes, the ODA's mission requires us to be outside in the elements, and this will require those preventative meds. Here's the issue though, when doc hands them out, these items never get placed in the individual's kit/ruck. This class of supply is not free to the unit. Each time the 18D has to stock up (in bulk) with these items, it limits his budget for other items.

Command and Signal. Good luck getting a termination station coordinated for every training concept. Do what you can to secure one for Collective training events but fight like hell for every FMP event. Your 18Es should be able to find an approved training SOI for your local training. Radios are a perishable skill. If you have a training SOI then incorporate radios every chance you get. At a minimum, the guys only turn them on for a circle of love/comms check prior to training events. No harm in an extra rep booting up the radio.

Actions after issuance of an OPORD. This will vary for each training event. Do yourself, more importantly your guys a favor and provide adequate time to prepare equipment, and or conduct rehearsals, PCIs, and PCCs.

PCIs/PCCs. Everyone has an opinion and experience with PCIs and PCCs. This is how I define both. PCCs are my tactical check lists for each MOS, specialty team, truck load-out, etc. PCIs are leadership spot checks of critical items (show stoppers), and the circle of trust inspection.

Circle of trust. If you have not heard of this term, it is a literal circle the team forms up in. The leader comes to the center and moves to his 2IC. The leader is inspected by his 2IC from head to toe, front to back, for all things mission related. After the men see the leader is squared away, the leader checks everyone else. I did not learn of this method until a few years post team time. We did however, a more informal version of this for as long as I can remember. Primary focus was radio checks, lasers, lights, and NODs.

Even if you executed an OPORD (day of, or day prior), be prepared to conduct that truck-side convoy brief prior to rolling out. This is for last minute changes that could push adjustments down onto the team. Adjustments could include, but not be limited to:

- PRI and/or ALT routes
- Initiation of the bump plan (number of aircraft/vehicles changed)
- Med plan for the route
- Hit time
- Crypto change over
- Intel dumps
- Task Org swaps

This training tool is only as good as you make it. Again, you cannot supply information for every line of the OPORD for a flat range day, but the idea is to utilize the format. The more you use it, the more familiar and comfortable your guys will become with it. As time goes by your ODA's planning and briefing products for future events, and real-world operations will become high speed, and low drag.

3-10.9 Now that I have finished my OPORD inspiration... Back to Training Concept tips. **Fire support**, the Captain and every SF NCO needs to maintain this skill, especially if your team is slotted against a mission where you will execute this task. 10th Group had a fantastic, all digital, training room for this. The Battalion fires officer can be a good source of information when trying to nail down this training.

3-10.10 **Communications base station**. This topic keeps showing up, but that is by design... collective training should have a base station to transmit to. Think B-TEAM, or Battalion S6. Maybe the GSB commo section has a system already set up, and ready to receive OPSKEDs and daily SITREPs. Or like the days of old, ODAs would leave a buddy team behind for TOC watch (house mouse).

3-10.11 **Realistic Military Training (off post training)**. If your concept takes you off the military reservation, your training concept will be robust to say the least. Realistic Military Training (RMT), or off post training concepts can require a big signature, like the Group Commander, 1st SFC CG or even USASOC CG. Make sure you, the 180 and your Captain understand all of the 1st SFC(A) Regulation 350-1 requirements for these training concepts (pg. 6-9 in 350-1).

You will have to coordinate with civilians (land managers/owners, law enforcement, etc.) in order to conduct off post training. Civilians and law enforcement are willing to help us out, even bend over backwards to make our training happen. Be prepared to make and present a certificate of appreciation at a minimum. Range time, or including them in whatever training you are conducting... If you want civilians conducting any military training, it requires a concept that is approved at high levels of Command.

Special or long-term land use agreements are difficult, but not impossible to establish. Your Group land use manager should have a list of approved land use agreements to tap into. If you are interested in a location that is not on the approved list, sit down with your land manager at Battalion or Group. There will be paperwork for the unit, and more paperwork for the civilian, state or federal land managers. I am going to spare my fingers from typing out the specifics I delt with at 10th Group/Colorado but do your homework to see what is involved.

When conducting initial coordination never make promises, verbal or written, to anyone outside the organization when conducting initial planning. This includes, but not limited to:

- Equipment vendors
- Lodging
- Off the shelf training vendors
- National Forest Service
- Retired or recently transitioned buddies who now represent any of the before mentioned organizations

Every land manager, vendor, contractor, etc. operate differently. You will encounter similar situations for similar tasks, but most will have a new twist. These twists will produce questions you do not have answers to.

Depending on the questions you have, make sure to run them through the appropriate channels. During my time in the seat, and the other jobs afterwards, these planning considerations took me to all of the below listed personnel, or shops, and a few more.

- Company leadership
- Battalion/Group staff (S3/S4)
- Group S8
- Judge Advocate General (JAG).

3-10.12 **Medical plan**. Do not let your medic half-ass the med plan. He should know every hospital location and the trauma level rating for the area you are training in. He should know how long he can keep a patient stable with the equipment on hand, for potential injuries congruent with the training being conducted. Lastly, how he plans to move casualties (and the time required to move them) to the predetermined medical facilities. His plan needs to cover each phase of the trip (travel to, at training location, and return home).

Flight for life is a great asset, how to request them is different everywhere you go. Some places you call them direct, others require calling 911. Know before you go.

3-10.13 **Never be satisfied.**[8] How can it be done better? You, as the Team Sergeant, need to maintain a running AAR when executing training concepts. So many incredible ideas and solutions were lost because I did not

[8] Part of COL Glover Johns' Leadership Philosophy, see annex II

write them down. Obviously enlist the help of the Captain, Warrant, 18F, and the other senior guys.

When training is complete, conduct an AAR. If the team is conducting a multi-day/week event, maybe a daily hot wash (with scribe) is a better way to capture all of the improves.

You and members of the team need to execute AARs with your thick skin/big boy pants on. If you as the leader, or an individual messed something up during the training event, own it. Do not be the guy defending yourself and making excuses for what happened. Accept the criticism as a teaching point and strive to be better at that task.

On the *Pineland Underground Podcast*, episode 40, Kyle Lamb and crew discuss AARs (among other leadership topics). Their view is similar to what I wrote, but they also stated that <u>sustains</u> should not be part of the process. AARs are not designed to make you feel good and talk about what you did right. Focus on what went wrong, write it down, and the leadership ensures a solution is implemented during the next rep.

I agree with them regarding sustains within an AAR. However, if we as a team introduced a new TTP/SOP, I will ask point blank to the guys, "did it work?" If yes, then no further discussion was needed. If no, then back to the drawing board we went.

3-10.14 **1st Special Forces Command (A) regulation 350-1**. Before we cover training concept examples, here is what I consider essential reading within 350-1. At some point you should explore the regulation in its entirety, but these select chapters and paragraphs are a solid starting point.

1st Special Forces Command 350-1.[9] **Read and know**:

- Paragraph 1-9, Detachment Commander priorities
- Chapters
    - 3 Training Management (paragraphs 3-4 thru 3-7 for LRTC)
    - 5 Individual Training
    - 6 Collective Training
    - 7 Medical Sustainment training
    - 8 Training resources
    - 10 JCETs

Scan these paragraphs:

- 1-6, 1-7 and 1-8 Group, Battalion and Company Commander's priorities
- 4-5 Lessons learned reporting requirements
- Annexes for 350-1 Concept templates

I am 99.99% sure you will never find this regulation on the Army Publishing Directorate. Check the portal or ask the Battalion S3 for a copy.

---

[9] I referenced the 24 JAN 2019 version, paragraph and chapter numbers will most likely change with an updated regulation.

Training Concept Examples

3-11 Good intentions, bad execution. 0332, Cold weather Unconventional Warfare. Conducted in Gunnison County, Taylor Park, CO. It has been a few years since this event, but here is how we got there, and some lessons learned to the best of my memory.

The team pitched this concept to the Battalion Commander at a Semi-Annual Training brief. At the time the Battalion was not on the playbook for Afghanistan, and Iraq had been shut down. 10th Group was in full swing to get back to our roots, operating in cold weather and mountainous terrain. Additionally, the Annual Training Guidance called for a renewed focus on UW.

With the boss' blessing, the detachment leadership designed a training progression with two primary goals. One, be "T" for trained in UW. Two, proper preparation in order to shoot, move, communicate, medicate, and thrive in a below zero, mountainous environment. The ODA completed a long list of training in order to get ready for this event. See below for some highlights:

- Downhill skiing (bunny hill through backcountry travel in low visibility with NVGs and full kit)
- Backcountry travel (recognize/avoid avalanche terrain)
- Snow science
- Avalanche rescue

    - Downhill/backcountry, snow science, and avalanche were taught /certified by the National Ski Patrol
- Rescue rope systems (high and low angle)
- Snow caves/winter survivability training
- Long range patrolling, TTP/SOP development on skis and snowshoes
- Specific physical fitness through Group's THOR3 program
- MDMP and Unconventional Warfare review

The event we planned was two weeks long, with an accelerated timeline, replicating six months of action on the ground. Picture Robin Sage, just colder. Without help from the Company or Battalion, my 18F developed the situation, prepped and coached the Guerilla fighters, OPFOR, and other role players. He executed all of this coordination without me, the Captain, or anyone from the team. I wanted to be fully un-aware of his scenario, and how he was going to maneuver it against us. He also worked with a few landowners and the National Forrest service Rangers to develop targets.

Our AOB was at half strength due to deployments. Regardless, the Company Warrant (an old school Fort Devens guy) was fully on board. He put together a small AOB backside support team to help the ODA. They received our transmissions and doubled as auxiliary role players. The Guerilla force were GSB personnel.

I wanted to infill via static line parachute. Not just to jump, because jumping hurts. But to go through an inflight rig sequence, push bundles (back country ski touring sleds/skis), and exit an aircraft in flight with snowshoes attached to the jumper. As with any concept, we had to do most of the prep work. When we conducted our initial training site tour and coordination trip, we conducted a Drop Zone (DZ) survey. We learned that the DZ altitude (9,000 plus feet above ground level) was too high for a static line infill. I was less than pleased we had to adjust our infill plan.

Lesson learned, don't be married to an idea. You will find yourself trying to force something when one, or twenty less complex solutions are available. Our infiltration method became an auxiliary guided, snow machine "border" crossing, with a ski tour movement to our link up location.

Another painful lesson was learned on that cross-country movement. When your team has been reduced by 25% due to taskers, you need to decrease the team gear packing list. I was too rigid with the Robin Sage mindset that we absolutely needed two of each radio systems, and a ridiculous number of batteries. This was on top of a double basic load of ammunition, crew served weapons, food and water for forty-eight hours, additional food for the Guerillas, cold weather sleeping systems, and the list went on.

I did not listen to my guys who provided some dang good solutions to reduce the heavy ruck sacks and sleds. Options that were presented, and I vetoed were:

- 18Cs build bundles for post link up aerial resupply
- Pilot team establishes cache pre-infill
- Bundles during the airborne operation (if it was a go)/create cache after assembly area activities and security is set
- Auxiliary/underground transports, stores, and delivers our bulk logistical items when requested

This led to one of the worst infill events I have ever faced. The heavy ruck alone was not the issue, we were pretty fit as an ODA. The snow conditions were our downfall. The top layer of snow, (4-6 inches) was a solid crust. A man with a chest rack, weapon, and normal rucksack weight (towed on a sled behind him/proper weight distribution), could glide (on skis) or walk (snowshoes) across just fine. But with my packing list we punched through with every step. After crashing through, the tips of the skis would get stuck under the hard top layer of crust, making it almost impossible to lift your legs up to walk. Below the hard crust was what is best described as sugar snow, you cannot make a snowball with it, and it is like walking uphill in deep sand.

We did what we could to maintain momentum. We swapped the point man every 100 meters, put all the men towing sleds in the rear, as the men up front attempted to compact the trail for them. The whole time we were battling the decision of taking the easier route lower in the valley, or remaining higher in the tree line, and traveling in a tactical manner for the terrain we were on. With the looming danger of OPFOR, and the horrible snow conditions, our stress level was pinging.

I cannot remember the details, but at the rate we were moving we would not make our link up time. The team conducted a long halt, established security, prepared a leader's recon/link up party, dropped rucks (minus avalanche/survival items) and they moved out to the link up location. 0334 was serving as the pilot team for the scenario. After link up was complete, it was a huge shot to my ego to admit defeat to another Team Sergeant.

As for the rest of the training event, we faced a good number of Sage scenarios. My 18F had done a great job to provide a challenging lane that tested our capabilities, but nothing too outrageous. If we could do it again, I would not have changed much for our first ever UW event. Although I would have put more effort into developing the following:

- Area command role player(s)/interaction
- More auxiliary role players/interaction
- Larger Guerilla element, to force a larger logistical problem for the ODA

3-11.1 The next two concept ideas were some missed opportunities. If you (and the guys) think of something, go for it. Some ideas require more work than others, and a few ideas you have the guys will not like, but you know they need to be done. For all scenarios, if your gut is telling you to do it, walk the dog on it. As you learn the training concept process, you will realize how quick a plan can be created if the whole ODA is working together. Even if the boss tells you no, during this Quarterly/Semi-Annual Training brief, shelf the concept, and pull it down for the next brief.

First, survival and E&E refresher training. We never got to do it, but I wanted to set up a concept that focused on re-bluing the team on this skill. Not just the knowledge and practical exercise, but the planning aspect.

As a mountain team, it would have been fairly simple to include all of the field craft training into any of our mountain skills-based concepts. As for the E&E portion I would have had the 18F and 180 plan out the corridor and corresponding requirements. Then I would attach it to an alpine skills sustainment/validation event. This would be more realistic due to the team exerting a great deal of energy working at altitude, then being pushed into a long-range movement.

For the E&E portion, the team would move along a planned corridor, navigate to caches, build ground markings, hit communication windows, and get extracted. Preferably, explore how to gain approval to eat what we kill. Of course, this would require outside support to emplace caches, maybe UAV support for ground marking observation. Coordination with local and/or federal offices to support our efforts would be a must. They would play a critical role in order to coordinate with the landowners and potential communities. I even wanted to coordinate with the Sheriff's office, ask if they would bring out their tracking teams.

Second, everyone's favorite, MDMP. As we all know MDMP is a perishable skill. If I could run another team, I would incorporate it more often. Outside of OCONUS trips, and applicable training events, I would dedicate two weeks, once a year for training MDMP.

To do it right, I would get the ODA leadership down to SWCS, and sit in with the guys who teach MDMP at Robin Sage.[10] After gleaning what we needed, I would secure a classroom, and assign the Captain as the lead instructor. The team would review MDMP in its entirety in three to five days, then the following week do a 72–96-hour isolation block.

If you decide to add this training to your LRTC, be creative. Dig through the OPLANs for your AOR. Talk to the S2, and all those little offices with no windows. There may be a real-world situation (in your AOR) you can conduct MDMP against. Things have changed since I have been at the tactical level. The Groups might have MDMP scenarios for your team to work with already, ask around.

---

[10] See SWCS MDMP product in Annex I, this is the best breakdown of MDMP I have ever used.

Your guys will not be fans, but proficiency in MDMP is a must if you want to be competitive. The glory days of GWOT are mostly over, ODAs will be fighting for the top mission sets within their AORs again. Training MDMP is not sexy, and often neglected, but it is the right thing to do.

3-11.2 **SFOD-A 5135's executed Training Concepts**.

> (EL) Below I have listed the end products, which does not include all of the build-up necessary to reach the FMP events. These concepts were completed within three years, between two deployments, and two Joint Training Exercises. Although the following list seems ambitious, the previously discussed trackers and systems keep the team on track.

- 160th SOAR Helicopter Dunker Trainer
- Helocast Night FMP (with Regular Army as Partner Force)
  -Progression Helocast training from pool to open water-slick to combat equipment, day-night.
  -FMP
  - night fly-a-way from Ft. Campbell to Paris Landing State Park,
  - Helocast entire ODA (4x ODA members with Combat Dive equipment)

    - simulated injury on "SPLASH" which required a waterborne extraction utilizing the Stokes Litter (from the MH-47)
    - 39km boat movement
    - Employ scout combat dive Team (4x divers) to clear the BLS
    - Secure the BLS
    - call in remainder of the element (on boats)
    - conduct link up
- Simulated Deployment with Regular Army Partner Force
    - "Deployed" to Ft. Campbell Ranges to conduct FOB Operations
    - Partner Force utilized Remote Advise Assist and Virtual Accompany Kit (RAVAAK) systems for the ODA to battle track and advise
    - Trained and executed
        - Ambush
        - Raid (2x)
        - Base Attack
        - MASCAL
            - TST
            - Team Evasion. Team helped plan our EVASION but was not aware that we were going to execute, only the detachment leadership was aware.
- Mid-South Tactical Shooting Institute (probably the best tactical shooting course I have attended AND it's only 5 days)
- Long Range Desert Mobility Course (Utah)

- 2x days of vehicle recovery techniques and planning for long range movement (employed NSCV's, similar to vehicles that we utilized on deployments)
- 3x days of long-range desert mobility focusing on maintaining the vehicles and sustainment for longer duration desert operations
- Long Range Kayak Movement with Combat equipment (validate or proof of concept utilizing military kayaks)
  - Used military kayaks loaded with full combat equipment to last 3-5 days (approximately 800lbs of personnel, equipment, sustenance)
  - All Team members wore body armor
  - EW Enablers to identify our electronic signature and actively trying to pinpoint our location
  - 60km overall boat movement
  - Broken down into 20 km movements
  - 1st 20km utilizing all forms of electronic signature (to establish baseline)
  - HF Shot
  - 2nd 20km greatly reducing electronic signature (only employed PRC 152 when necessary)
  - HF Shot
  - 3rd km conducting split Team and night link up
- FIDEX with TAC-P's on Ft. Campbell (emphasis on conducting Raids)
- Progression of conducting raid operations
- 3x FMP Targets with SSE and intel on target
- Final raid was a fly-away from Ft. Campbell to Ft. Knox and return to Ft Campbell the following day.
- SUT at 5th RTB (Dahlonega)

- SUT with SL/CE INFIL
  - Dive Re-qual
  - Harbor INFIL (Jacksonville, FL)
  - FMP was a night Helocast 50km from harbor
  - Long range boat movement (zodiacs) to 1500m from harbor gate
  - Splash combat dive Teams to infiltrate the harbor
  - Combat dive Team place limpet mines on USS Battleship
  - Exfil dive back to boats (zodiacs)
  - Long range boat movement 35km to ocean link up with MH-47
  - "Scuttle" zodiacs (hand over to dive locker) and conduct waterborne exfil with caving ladders into the MH-47
  - Return to friendly lines, ASO (tie in)
  - Splash combat diver Teams to recover a sub-surface dead drop (intel)
  - Splash combat diver Teams to infiltrate a harbor and emplace a tracking device on a "target" boat (dive locker boat)
  - Splash combat divers to emplace a camera on harbor pillar
  - Track "target" boat to location and conduct a raid
  - VBSS with U.S. Coast Guard
- Fast Rope INFIL with MH/AH-6 Little Bird (160th) for Raids at Mining Facility (West Virginia)
- Ground Force Commander Course (at the Group JTAC locker)
  - Employed the entire ODA, not just Team leadership

  - 5x guys in OPCEN (TL, TM SGT, PF Liaison, Front Line Trace, JTAC)
  - Remainder of the ODA at field with MFCS and 2x 81mm mortar tubes
  - Worked the communication piece with fires de-confliction
- Dive at Ripley's Aquarium (to dive with sea creatures)
  - Controlled, salt-water environment with live creatures
  - Buoyancy validation for both open/closed circuit equipment prior to dive re-qual
  - **RECRUITMENT**
- Dive and MFF Team Joint Operation (planned but did not execute due to PCS)
  - Night, Fly-away from Ft. Campbell to Paris Landing State Park with both specialty ODA's (1x Team per A/C)
  - INFIL dive Team by Helocast (to "secure" the area, but also provide safety boat coverage for the in-bound MFF jump)
  - Once "secured" MFF Team conduct INFIL
  - Link-up with both ODA's
  - Long Range boat movement with both detachments
  - Arrive to "splash" location
  - Conduct a scout combat dive to secure BLS
  - Once BLS Secured, bring in the remainder of the element
  - Reach ORP, and establish MSS
  - MFF/SR Team pushes forward to RECCE the objective and provide products to the Dive/DA Team

  - MFF/SR Team provides Isolation and B-TEAM provides containment (with trucks)
  - Dive Team conducts Raid
  - EXFIL together with B-TEAM support
- Internal PMT
  - Executed prior to EXEVAL or PMT
  - CONOP at Team Room
  - Patrol to Range with USASOC IED rep emplacing simulated IED's en route
  - Once at Range, it was standard training, range conduct
  - Upon departure of Range, USASOC IED rep emplaced simulated IED's on exfil
  - Maneuver live fire with IED reaction
  - Call For Fire/Mortar (adjustments with drone-observation)
  - Raid
  - TST Raid (Hasty Planning)
  - React to IED/SVEST
  - MASCAL
  - C-UAS battle drills with attached EOD assets

3-12 To close this chapter out I would encourage the reader to review a few of CSM Dorsh's Notes on leadership:

> **5) Analyze your training calendar**. What are you putting your resources against (time/$)? Are we devoting time/$ to personal development of our up-and-coming leaders (you)? We say we are doing it, but are we? Your calendar will tell me if

you are. We have train/man/equip responsibilities while here at Carson/Panzer. Leader Development of our personnel is part of that. We tend to focus on operations – because that's easy – that's what we know. Training leadership is hard – find a creative way to do it.

6) Fight complacency. It's human nature to take the path of least resistance. Strive to get better every day. Put yourself in uncomfortable situations. Force yourself to learn something new every day.

10) Training – Always work to incorporate a psychological and physical component to training. **Work to inoculate stress into your formations.**

11) Competition Breeds Success – Man in the Arena - Tomorrow is not a right; Earn it! Those are my mantras – What are yours? **Does your video match your audio?** You are moving into the realm where everybody is watching what you do and listening to what you say. Be conscious of that.

These four Notes strike directly at the core information covered in this chapter. Be sure to incorporate them into your training development process. Doing so should keep the ODA on the boss' radar for the good missions.

4

On OCONUS Training

4-1 Some JCETs can feel like a paid vacation, just one of the perks of the job. Although that may be true, use these trips as a full-dress rehearsal for the team. Evaluate your guys, and how all your checklists, systems, training methodologies, and POIs work together.

The boss and powers to be only have so many data points to measure and compare ODAs. We know they pick the best ODA(s) for the important missions. With that knowledge, treat every TSCP event (JCET, JPAT, CNT, etc.) like a trip to war. I'll keep saying it, use that TS and dig into your Group's AOR. You (and the Captain) absolutely must know why you are traveling to random country X. Your understanding of the why will help the Captain develop the Mission statement, his Commanders' intent, Expanded Purpose, Key Tasks, and End state.[1]

4-2 Modified MDMP. This is a perfect opportunity to practice MDMP within the 180-day timeline. You knock out IPB and start MA, then pause. After the PDSS continue through the rest of the steps. These missions are "easy" to plan, so products should not be ridiculous to

[1] See Annex I, pages 318-319 for additional information on Commander's Intent, expanded purpose, key tasks, and end state.

create. Conducting MDMP for a JCET also helps build future Isolation templates.

The JCET timeline is no joke, if you are in the 180-day countdown you should be reviewing the product checklist[2] on the first and last day of each week. You must apply pressure to the guys. Ensure they know what they are responsible for by assigning names to specific tasks.

If you know that you have an overseas trip coming up, and plan to utilize MDMP, make the time to knock out some refresher training. If you go into the planning cycle half-cocked, the team and leadership will become frustrated, which will end in a subpar product or, none at all.

4-2.1 **Product submission**. Do not be surprised at the number of times you are told to revise and resubmit documents. In conjunction with the last point, another fun task is making sure Battalion doesn't drop the ball with your products. You may work 20 hours a day, but staff guys do not.

If you do not have "permission" to interact with Battalion staff, use the weekly Company training meeting as your platform of inquiry. Be the squeaky wheel that gets the grease.

4.2.2 **IPB**. Chances are high that you are not the first team to travel to your assigned vacation destination. Read every post mission product you can prior to going on the PDSS,

---

[2] See your BN JCET manager for a current copy of the product checklist.

don't look dumb asking questions that have been answered. These products should answer most of your RFIs but verify everything on the PDSS. Also, this will help when developing your training plan. Reach out to the unit Lessons Learned Contractors if you do not know where the post mission documents reside.

The 18F should build a robust country study for every country you deploy to. He should have access to products that only need to be updated, no need to recreate the wheel. Whether it is a digital or paper product, the team needs to review it like their life depends on it. Your Team shouldn't go into a country without knowing everything about it.

We are supposed to be the masters of the human domain. Our purpose in life, combat multipliers, relies on our ability to bond with our Host nation/partner force. In order to bond, you need the aptitude to dig deep into their concerns, demonstrate you know their customs, ask about their history, and the country's accomplishments.

Better said by Hy S. Rothstein in his book *Afghanistan & the troubled future of Unconventional Warfare*:

> SF must be diplomats, doctors, spies, cultural anthropologists, and good friends-all before their primary works come into play.[3]

---

3 Hy S. Rothstein, Afghanistan & the troubled future of Unconventional Warfare, chapter 5, page 144

The best springboard, for the above concerns, is the PMESSI/ASCOPE cross walk. A deep dive should define the Host nation disposition, which will identify knowledge gaps we need to fill. When you travel through the TSOCs for your authorities and permission briefs, be sure to confirm or deny what you developed. Depending on where you are going, the country's, or the region's military history may be a source of pride. Or, maybe a topic to be avoided at dinner. Be sure to dig into the area's history.

More importantly the IPB process will help identify real world security threats within the Enemy Forces section:

- Composition
- Disposition
- Strength
- recent activities/SIGACTs
- suspected locations/capabilities
- MLCOA/MDCOA.

Use this information to assist in preparation of the team's Threat Vulnerability Assessment.

4-2.3 **Go to hell plan**. If you developed a TVA, part of that plan answers the question "what happens if we cannot stand our ground?" Have the 18F and 180A develop an E+E plan. Try and make it realistic to your no-shit situation. Do not forget the basics, everyone on the Team should know how and when the plan is activated.

Every Team house requires a destruction and bug out plan. No matter where we are in the world, there is someone who does not like us, and may want us to leave in a hurry.

4-2.4 **OPCEN**. If you have an OPCEN, does it need 24-hour manning? Every situation is different, use common sense. Be ready for visitors that are not cleared to see your stuff. Hang rolled up sniper screen, ponchos, or tarps over white boards, maps, charts, base defense diagrams, etc.

4-2.5 **Training plans**, have a few on deck. Be prepared to conduct a thorough assessment of the Host nation/Partner force. You might be surprised at the skill they demonstrate. Do not be the Team that comes in with a basic ass plan if these guys are flowing through the house like pros.

4-2.6 **FOO / PA**, pick good dudes, have alternates. Each AOR is different, know what qualifications are needed, where they go for the briefs and classes. Know where and when they can draw funds, plan accordingly to have them at the right place and time so there are no delays once you hit the ground.

4-2.7 **Money**. 18A, 180A, 18Z, and 18Cs should know all of the pots of money. Once you know what each pot does, you can build realistic OCONUS budgets, equipment lists, training requests, etc. You need to be the commonsense guy when it comes to equipment orders. Know the difference between mission critical, essential, and enhancing items. The more you know, the better chance of acquiring the right equipment for your mission.

4-2.8 **Post mission products**. I am going to remain vague here for OPSEC purposes, but here are some tips to produce quality post mission products. Your Battalion JCET manager will have a list of required products your team needs to complete and turn in after your trip.

Leave post mission products open on the desktop in the OPCEN for the whole trip (common sense, save often/save always, and back the work up on disk daily). Have anyone that interacted with the Host nation fill in data every day. This will do a few things:

1. All ODA members contribute

2. The product is basically done prior to heading home

3. An actual account of what happened on the trip is captured

4. NCOPD for your NCOs who will take a team someday

4-2.9 **Cultural exchange**. Be prepared to bring or create awards for close out ceremonies. A paper certificate with GB flashy stuff on it goes a long way in certain countries. Bring SF swag (SF patches, Team patches, US flags, team shirts, stickers, coins, etc.) for trade with the Host nation. Some countries even have local places to create plaques. Otherwise, if you have the required unit information you can have something created stateside and bring it with you.

4-2.10 **Medical plan**. Do not let your medic half-ass the med plan... a pattern has emerged! He should know:

- Allergies, ensures each SM draws required medications before the deployment.
- All other prescription meds for each team member.
- Will Your 18D be the primary sick-call doc for your trip? If so, he should dig into what he needs to bring, and possibly set up a resupply plan depending on the trip length.
- How long he can sustain a casualty with equipment on hand (field vs. team house).
- Location of every major hospital, at least the ones close to your area of operation.
- The trauma level rating for those hospitals.
- Have a plan to move casualties to those hospitals for every phase of the trip-
    - Travel to country X
    - At training locations
    - Travel home

If possible, rehearse the med plan. Another consideration is how does the team send a guy home if he is injured? Every TSOC should have something in place for this scenario. Lastly, your 18D (if required), be prepared to conduct an evaluation of the host nation's medical facilities while in country.

On my last trip as a Team Sergeant the ODA traveled to a destination hot spot in Africa. The training location and team house were located in a small town that was a three-day drive from the capital city. Our location was austere to say the least, only made better by the uranium mines that surrounded us.

We were co-located with a fantastic Foreign SOF team, who we shared training objectives with. We also shared a MEDEVAC platform. Understanding our predicament, both teams came together early on and developed a joint SOP to handle a few medical scenarios.

Our Medics obviously focused on patient care, packaging, and transportation. The 18Bs, foreign SOF tacticians, and the team's leadership focused on routes, danger areas, convoy task organization, sustainment, and command and control measure issues. Once the plan was completed, the leaders agreed we needed a full rehearsal.

When the agreed day came (only the leadership for both teams knew which day), both teams initiated the rehearsal with a few casualties during a simulated probing attack. Both teams locked down security, and the medics began their care and packaging protocols. My ODA was further away from the airport, so we collapsed our position and enacted our Go to hell plan, to include "calling in" the 9-line.[4] The plan called for a link up at the foreign SOF house, loading of critical gear and patients, then depart for the airfield along the planned route.

The event ended when the aircraft arrived, and the medics finished with the handover brief with the aircrew. Although the rehearsal revealed many areas we needed to tighten up, it demonstrated to both teams we were on the right path. Goes without saying, I left out a whole mess load of details in an effort to maintain some resemblance of OPSEC.

4-2.11 Not every trip we go on is a combat rotation. Get out there and see something new. Obviously, only if the real-world threat situation supports it. The TSOC, country team, and your 18F should be dialed in and will let you know if this is a possibility.

You should have consumed enough booze in your younger days to stay out of the bar... at least long enough to do some site seeing. Inspire your guys to learn some of the historical aspects of the country you are in. This ties into that cultural understanding we are supposed to be good at. Your 18F should have identified five or more different things to go and see while developing the country study.

---

[4] Our MEDEVAC platform was a contracted aircraft that doubled as our logistics carrier. For this event we had a pre-planned logistics run to be brought in.

5

On Leadership

***"A man must be big enough to admit his mistakes, smart enough to profit from them, and strong enough to correct them."* – John Maxwell**[1]

5-1 The above quote sets the tone for this chapter. No one in our profession likes to be wrong, but the Good and Great leaders I have worked for always adhered to the above quote.

As I mentioned in Chapter 1, there are plenty of quality leadership books out there. The goal of this chapter is to share my Monday morning quarterback/if I could do it again insights. I have leaned heavily on both Paul Lefavor's and Kyle Lamb's work throughout my book, why stop now...

5-2 First up for this chapter is - know yourself before you lead others. If I could give my younger self any books before taking a team they would be: *US Army SUT Handbook, Tactical Leadership, and Leadership in the Shadows*. Below is why I believe these selections are a good place to start.

---

[1] John C. Maxwell (2001). "The Power of Leadership", p.18

First up, *US Army SUT Handbook.* Lefavor's "Leadership" chapter is not just a checklist one should follow, but a well-researched product that explores a few important topics:

- What makes a successful SF Soldier?
- The Twelve Leadership Traits
- Situational Leadership
- Officer/NCO Professional relationship
- The Eleven principles of leadership

This information induces critical thinking towards one's abilities in the art of leadership. In my opinion (after learning the hard way) knowing yourself, in regards to the art of leadership, is just as critical as knowing where you are on the map during an operation. If you don't know your location, then you cannot navigate to your target, call for fire, call in a MEDEVAC, or even return fire for fear of fratricide.

Conversely, if you do not know your own temperament (a combination of inherited traits which subconsciously affects our behavior-SUT Handbook, Ch 5, pg. 218), you do not know your own strengths and weaknesses. This makes navigating our specialized social construct more than difficult.

Understanding your own tendencies, temperaments, strengths and weaknesses can smooth the path before you. Example of this is when faced with tough times, I reverted back to what I knew, which was dictatorship in nature. As I described earlier in chapter one, the "my way or the highway" attitude, and micro-managing team guys is not recommended.

Don't be like me, finding yourself halfway through your Team Sergeant time and asking, "Why are the guys going to the Captain for things like appointments, passes, and leave?" Or, "Why are my senior guys volunteering to go to 1/10, or another team?"

Do yourself, more importantly, your guys a favor. Conduct self-assessments constantly. If you identify an area you want to improve, do it. Continue to read everything you can on the art of leadership.

Speaking of reading, Lefavor's "Leadership" chapter is well laid out, and specifically geared to our Green Beret way of life. In an effort not to quote the whole chapter, here are two things that hit me like a lightning bolt:

1. When I reviewed his temperament quad-chart I learned that I aligned with Sanguine, and had a touch of Melancholic tendency:
    - Sanguine - People oriented, Extrovert
        - Sociable pleasure seeker
        - Charismatic, creative-daydreamer
        - Prefer to do things as a Group
        - Flighty, may have trouble following through with tasks
    - Melancholic – Task oriented introvert
        - Tend to procrastinate and plan in excess

2. In Lefavor's chapter conclusion paragraph[2], he poses the following questions:

> Looking back at this chapter and evaluating our own leadership, how well do we stand up?

- Are you able to use the ideas of others?
- Can you accept criticism of your decisions(s) without taking offense?
- Do you depend on the praise of others to keep you going?
- Can you hold steady in the face of disapproval and even temporary loss of confidence?
- The longer you lead, do things improve or get worse?
- Can you anticipate how your words will be received?
- Can you forgive? Do you nurse resentments?
- Do you criticize or encourage?
- Can you admit it when someone else's view is better?
- Do you make excuses for your failures?

Next in the shoot, *Tactical Leadership*. After reading the leadership chapter in Paul's *US Army SUT Handbook*, I

[2] Page 231, Chapter 5 Leadership, "*US Army Small Unit Tactics Handbook*".

remember thinking - this should be its own book... well, sure enough it is! In *Tactical Leadership,* Paul is able to conduct a deeper dive into all of the topics, and more covered in the *US Army SUT Handbook's* "Leadership" chapter.

What really struck home for me in *Tactical Leadership* is he provides 'theoretical leadership experience' by reaching back in time to past military leaders. As he reviews a campaign, or a battle he uses a specific lens, for example, Tenacity has its own chapter. In it he visits actions from John Mosby (the Gray Ghost during our Civil War), and SFOD-A 765's leadership in the early years of Afghanistan.

The full list of specific lenses he uses are:

- Intellect
- Imagination
- Judgment
- Initiative
- Decisiveness
- Flexibility
- Determination
- Courage
- Perseverance
- Tenacity
- Self-control
- Presence

After Paul recounts one, or a few battles for each leader, he summarizes the critical leadership lessons learned with the acronym LEAD.

- L- Limitations to avoid
- E- Examples to Follow
- A- Action to Emulate
- D- Doctrine to understand.

Paul's objective approach helped me by providing context to each Leadership trait / skill.

My last book recommendation on the journey to better understand yourself is *Leadership In The Shadows*. After getting to know your tendencies, having a deeper understanding of the 12 Leadership Principles and supporting traits, you should set your sights on what type of leader you want to be.

Kyle Lamb breaks down this topic into five different categories. I have provided a few of the key points for each leader type that he identified in his book, but it goes without saying, you need to get his book for the full takeaway.

- Great Leaders
    - Real world experiences to make great decisions
    - Have heart/spine to make hard decisions
    - Foresight to predict second and third order effects of their decisions

    - Have earned credibility and respect
    - Take care of their people
    - Not driven by promotion, but driven by mission accomplishment
- Good Leaders
    - Constantly working to:
        - put the men first
        - take responsibility of their actions
        - establish their credibility
    - Lacking time and/or life experience in current position to be a great leader right now
- Bad Leaders
    - Physically present, but not fully committed to the men or the mission
    - Lack the decision-making process experiences to make sound decisions in a fluid situation
    - Like to be "one of the guys," when they need to be their own person
    - They come and go without making an impact, *inconsequential*
- Dangerous Leaders
    - Have been told by leaders that they are great, and believe what they have been told
    - Make uneducated, poorly thought-out, life-threatening decisions
    - Will lose, or never had any credibility with subordinates

- Malicious Leaders
    - 100% in it for themselves
    - No interest in the mission, or the men
    - Using this position as a steppingstone to the next one... check a box
    - Never be accountable for their actions, or actions of their men

Knowing what right looks like - Great and Good leaders - and what to avoid - Bad, Dangerous, and Malicious leaders - is a fantastic place to start your Team Sergeant time. If you can look at yourself in the mirror each night and know you lived in the Good and Great leader columns, you should be able to sleep well. As mentioned in Chapter 1, you will have bad days. Keep this list of traits handy. Maybe hang it on the wall next to your computer, review daily, and execute how you want to lead.

Bottom line, if I had the above information prior to/during my time in the seat I could have combated my negative tendencies with the positive. Learned how to use other's strengths to help balance out my weaknesses. By understanding myself, I could have understood the men better. Lastly, If I knew what kind of leader I wanted to be, I would have had a list to measure my actions against daily.

Whether you read the books I have mentioned, or any other Leadership works, be open to what is said. Absorb what is relevant to you and apply it to YOUR personality.

5-3 Once you understand yourself, now you can focus on the men.

Leadership principle 4: Know your men and look out for their well-being.

5-3.1 There are way more personalities on the teams than covered in this book. I selected the following personalities because I grew the most as a Leader when faced with these types of individuals and the situations we found ourselves in.

5-3.2 **Strong will.** During your team time, prior to being promoted, you will have worked with many strong-willed personalities. Our profession is a beacon of hope to those who want to be surrounded by like-minded and physically capable individuals. Some of those guys were fantastic to interact with, they most likely served as your direct competition on the team. Or these individuals pushed you and the others to be the best version of yourselves physically and mentally.

Looking back (or currently) did you have that super savvy, super fit, but older guy? Maybe he joined up later in life, has a college degree and ran a business prior? He was well spoken, and everyone in the chain of command knows, and likes him. Although he is a wonderful asset to have on the detachment, some believe this kind of guy to be a leadership challenge, I disagree. Looking back a few paragraphs I concluded I should have used my team's strengths to balance out my weaker tendencies. These are probably the brothers you can lean on for that. The worry of them taking over or establishing a shadow government is low if you are dedicated to building and maintaining Leadership Currency.

Then there was that other guy. Looking back, do you remember that dude on your team? He was the one who received a lot of attention from the Team Sergeant... not in a good way, but more in the form of FM 22-102, Wall to wall counseling. Funny enough, after some TLC from the Team Sergeant, he was a solid team player.

Although I had a few of these individuals, the one that comes to mind now as I am writing was during my B-TEAM Ops Sergeant time. I had a very motivated 18B, who had been assigned to the B-TEAM for an extended stay. This was due to a few of his own mishaps, also his personality was not the easiest to get along with.

Knowing this going in, I found that giving him a clear purpose, well defined right and left limits, and always following up proved to be all he needed to come into his own. He turned himself around, started developing positive relationships within the Company, and before I departed the B-TEAM, he had earned a strong NCOER.

The important part of this story comes in about a year later. I was now the Team Sergeant on 0332. I received a call from my former 18B's current Team Sergeant in 1/10, regarding my rating of this individual. I explained the reasons above as why this individual was rated well. He enlightened me on a few of the things my former 18B had recently done, which induced the following leadership epiphany.

My method worked in a B-TEAM setting, a fairly controlled environment considering the scope of our profession. An ODA does not always allow for 24-hour daycare activities. I realized then, if I ever have another one of these individuals on my ODA, I would have to incorporate this lesson learned into my task organization calculus. Some guys might be good at their job, and work well within the team's culture, but they can never be tasked with specific work that requires little to no oversight.

You will have one of these guys, if not a few of them during your tenure. These guys are separate from the "every ODA needs a pirate" individuals. Although you will spend time working with the "pirate," you will spend significantly more time with these guys. I was told once, regarding these individuals:

> "If you let your boot up from their throat, they will immediately sprint to their own destruction, and possibly ruin the ODA's reputation."

5-3.3 **Harmony.** As a Team Sergeant, you must maintain the harmony on the ODA. The days of you navigating these issues as one of the boys are over. Everyone will look to you to handle all the personality conflicts on the team.

Looking back to Chapter 1 I stated:

> In fact, I can say without hesitation, at times you must be prepared for adult daycare activities.
>
> Being a Team Sergeant involves managing personalities and employing (task organize) the right guys with the right Teammate, on the right mission, or objective.

You must set the tone early before disagreements become conflicts. Early is the key word here, just like bad news, these personality conflicts do not get better with time. There are more than a few approaches out there, and depending on the environment (location, training event, combat, etc.) you need to intervein accordingly.

Examples: What if the team is local and conducting training? A disagreement between two guys becomes distracting, and you do not have time to stop to deal with the issue. What worked for me was to make a quick task org shift, so they have minimal interaction. Sure, you could give an order for those two not to speak until we get back to the team room... but let's be honest, depending on the personalities, one of them will feel the need to get the last word in.

Once you are back in the team room, find a spot to speak to them without an audience. You can talk to them individually, get each side of the story. Then bring them together to provide tools so they can hash it out. Or you can sit them both down and moderate a conversation. For either option, you should remain as calm as possible, ask questions to get to the heart of the matter. Once you identify the issue, reach back to your experience and talk them through how you handled something similar, or how you have seen others do so.

What if you are on a multi-day training event, or a real-world operation? You need to interject immediately when these disagreements happen, and quickly remind them of the mission at hand. Letting the guys argue in front of your partner force, students, etc. can and will degrade your team's credibility. Adjust their head space back to what the team is doing and remind them that the mission is bigger than their issue(s). Here, I found that giving them tasks to focus their attention helped them get through the event, and back to the team house. A task org shift may be necessary as well.

Once the event is complete and the team is back conducting post mission activities be prepared to execute your method of getting to the heart of the situation. Begin to apply the requisite amount of coaching and/or mentoring to regain harmony.

At the end of the day not everyone will get along. Look back at Chapter 2, **Not every Green Beret (GB) is a good fit for your ODA**. You are required to do all that you can to make the team function. You must find a way for the guys to understand this. Get them to a point where they can effectively work together and know that they are not required to be best friends. Remember there is not an overabundance of extra guys on the B-TEAM to make hot swaps.

You may run into issues with your Captain, or 180. Whether these matters are simple or complex, do what you can to handle these issues at your level. Don't forget to clear out the team room before the "conversation" begins.

How you approach these conversations is important. Yes, it is your team, but you are a member of the United States Army. You swore an oath to "obey the orders of the President of the United States and the orders of the officers appointed over me." You need to conduct yourself in a manner that respects this oath, no matter how mad you may be.

I just recently heard this term, and I think it is brilliant. "Communication transmission."[3] Everyone you interact with gets a different speed and/or intensity of verbal communication.

Examples: When you were a private, you had two speeds. 1st gear was for your peers, always smoking and joking. 2nd gear was for everyone who outranked you.

As a buck sergeant your transmission had three or four speeds. 1st gear is for your peers where you are free to be

---

3 CSM (Ret) William "Bill" Hanes.

yourself, tell inappropriate jokes, and gab about anything important to you. 2nd gear is for talking to everyone who outranks you, but now you have a professional tone. Even with the professional tone, it is not polished, in fact it is blunt. 3rd gear, this is reserved for those in your charge. As a young leader you are just beginning to find your leadership style. Your temperament is not fully developed, and you may yell more than needed. The possible 4th gear was reserved for your wife, and that gear could have had a few sub gears depending on the situation at home.

Now, as a Master Sergeant you will have many gears to select from. Depending on the path that brought you here will determine the breadth of those gears. Some gears can be used for multiple engagements. For example, and I am not saying this in a demeaning way, but the gear you use with your kids can be shared when interacting with the Captain.

Think about it, the Captain is new to this life. He has brought a fresh set of eyes and energy to the team and wants to make a big impact. Sound familiar to the dads reading this book? If you bring the hammer down in every interaction, or shoot his ideas down each time, what is that going to do to his confidence?

At the end of the day, not all situations can be fixed at the detachment level. Once you find yourself at the limit of your abilities, before the pressure cooker explodes, be sure to utilize your AOB leadership.

Going to the Company leadership is the same concept as your team guys failing to work out issues amongst themselves, and bringing it to you to fix. You are able to handle their situation because of experience growing up on a team. Now the AOB is providing a new perspective and level of experience to your aid.

Not only are they tapping into their own personal experience, they have a new peer Group which has a larger aperture. This wide-angle lens has seen and delt with more leadership issues than you have. They will have a different approach for your issue.

The other side of the above paragraphs is having a great Captain. If you have a Captain that does his job well, is physically fit, and does not mess around in NCO business, then consider yourself lucky. I was lucky, but I failed to help him when he needed it. We both strongly believed in work hard, play hard. This was fine until it came to that point in the night where an adult had to shut it down and herd the cats.

Had I done my job, and been the voice of reason, he could have avoided some backlash from our boss. I did not protect my Captain, who absolutely deserved it, and that is another huge haunt I have from my time.

You will know when it is time to step up in the above situation, or anything else you find yourself doing. Whether your gut, or common sense is pinging, listen to it. You know the rules of the road, and you have heard all the stories floating around Group. You know what happened to those who got caught dancing in the gray area. Take care of your team, one bad event can, and will affect the ODA for your time in the seat, and possibly for years to come.

Leadership Principle(s) 8 and 9: Ensure each task is understood, supervised, and accomplished / Develop a sense of responsibility in your subordinates.

5-3.4 As you get to know how the team operates, who works well with whom, now is the time to build one of the most important facets on a team...Trust.

To earn trust, you must give trust. As a Team Sergeant you do not have time to be balls deep in the weeds. Your responsibility now is to teach, coach, train, mentor and manage the team. All while maintaining a full situational awareness of every aspect of the ODA.

To build trust start small, just like you would with your guerrilla or partner force. Give them confidence tasks and targets, set right and left limits, with a clear backstop. Provide realistic completion times, and of course supervise. If they fail the task, ask yourself, did you give them concise guidance? Did you provide all the tools required to complete the task? If you did, use it as a teaching opportunity for him (fail early, often, and forward).

As tasks become more difficult and/or complex, set intermediate check in dates, conduct an azimuth check, and make adjustments. If you allow creative freedom, your guys will take ownership of the task. If they have ownership, then they will become fully invested and give the team their best efforts.

As mentioned, a few times now, not everyone is suited to be a Green Beret. The above can be another metric regarding a decision to keep or remove a guy from the team. If you have done the above, and all aspects of this topic covered in Chapter 2, it might be time to replace the SM in question.

5-3.5 As trust begins to build, you will identify your pipe hitters. I considered my pipe hitters to be those on the team that knew their job inside and out, consistently sought out self-improvement, and were always helping others on the team. Do not burn them out. You will have one, maybe a few NCOs that can complete every task on time, to standard or better every time. We as leaders discover this guy, then we lean on them. Or these individuals will voluntarily take on too many tasks which can, and will lead to burnout. Most of these individuals will continue to grind, but something in their life will suffer. Ensure they know when to speak up and tell you their plate is full.

5-4 Be dedicated. If you have been selected to run a team, then being dedicated should be one of your core attributes. This job comes with an awesome volume of responsibility, and at times can weigh on you to the point of wanting to throw your hands up and walk away. The amount of work you and the guys put into developing - training concepts, preparing for JCETs, requesting specialized equipment, etc. can, and will be ruined by some new policy, a re-prioritization of funds, or anything else that can scrap months of hard work. The team will look to you in these situations.

When these things happen always remember we are a bottom-up driven organization. If you let the BS defeat you, then you will lose focus on this fact. No matter what derails the team, take the time to re-evaluate your LRTC and re-work it so it remains nested with the Annual Training Guidance, strategic level vision, and that follows the Marlow product. If the LRTC changes, update your 6–9-week calendars and the supporting detailed training concepts.

Again, the Battalion Commander has limited time to get to know your team. You and your Detachment Commander will not get that many opportunities to build the trust required to reduce micromanaging. If you are dedicated to the fundamentals in this book, and can navigate the BS, your team's ability to adapt will speak volumes for your team.

5-5 The last major focal point I would like to cover is, bad news does not get better with time. Even the best teams have a bad day, when things go wrong, and they will, be prepared to inform the boss. The first step is ensuring you and the Captain have all of the relevant data before making the call. Generally, the 5Ws are enough to get the process started, but be prepared for the countless number of RFIs that will come your way. Most bad scenarios are covered under the published CCIRs. Dependent on the training being conducted, having a copy of the CCIRs in your (or the Captain's) leader book, range book, products that go forward with you, is a good practice.

5-6 OK, now that I laid out my important leadership takeaways, what is the glue that binds it all together?

In my opinion, Leadership Currency is that glue. In Chapter 1 I briefly mentioned this topic, we can now go deeper into Leadership currency. Since Chapter 1 was a few pages ago, lets revisit what I said:

> If a leader weighs every action and decision against these four "Cores", their leadership currency will continue to grow. When any one of these is not factored into your actions, this is when your reputation begins to decline.

Below I included a deeper look into Covey's[4] four "Cores" of Credibility. I did my best to relate the supporting information to life on an ODA.

---

[4] Steven M.R. Covey, The 4 Cores of Credibility, resources.franklincovey.com

Character:

1. Integrity requires these additional qualities: congruence, humility, and courage. He closes Integrity with this question: "So how do we go about increasing our integrity? Make, and keep commitments to yourself. Stand for something. Be open."
2. Intent, which he includes: Motive ("examine and redefine your motives"), Agenda ("Declare your intent"), and Behavior ("Choose abundance. Abundance means there is enough for everybody").

Competence:

3. Capabilities, "One way to think about the various dimensions of capabilities is to use the acronym TASKS (Talents, Attitudes, Skills, Knowledge, Style)."

- TASKS acronym breakdown:
    - "*Talents* are our natural gifts and strengths.
    - *Attitudes* represent our paradigms – our way of seeing – as well as our ways of being.
    - *Skills* are our proficiencies, the things we can do well.
    - *Knowledge* represents our learning, insight, understanding, and awareness.

    - *Style* represents our approach and personality."
- Covey suggests the following three items to assist with increasing capabilities: "Run with your strengths, keep yourself relevant, and know where you're going."

4. Results, "They matter!" Covey is correct here, you can have all the good intentions you want, but at the end of the day did your team get across the finish line? Covey lists three things to focus on to improve results: "Take responsibility for results, expect to win, and finish strong." See below for my takeaways:

- "Take responsibility for results." This does not mean you claim all the glory when the team executes well. You market the team's accomplishments up the food chain, secure the next top mission, write those awards, or fight for additional schools. On the other hand, when the team falls flat, you toe the line, and take the blame. Taking the blame does not mean sitting around sulking. Your job is to analyze what happened and develop a strategy to get the team across the finish line, or to do it better next time. The men need to see the right way to handle adversity, some of them most likely never saw it growing up.

- "Expect to win," not a foreign concept to us, but... keeping the team's momentum moving forward when everything is working in your favor is easy. When the chips are down though, the team looks to you. Set the example, look for solutions to achieve your goals and ultimately complete the mission. Sure, you may have to trim some fat from the overall end state, but fight like hell to complete the task, training event, or mission you started.
- "Finish strong" can apply to many things on the ODA. One example, a team will jump head-first into preparation for a training event, or a trip overseas. As time goes by though (sometimes months), this event gets pushed to the back burner. You must be the one that keeps the guys engaged with ALL products and projects. This directly ties back to Chapter 2 with **Battle Rhythm**.

  More on "Finish Strong"; what we do is inherently difficult. When the miles are never ending and the ruck is heavy, your guys will let you know. They will come to you with 10 different ways to make the task "easier." Sometimes they are 100% correct, and you can implement their ideas. Other times, they are just tired and want the pain to stop. Stay strong and show them how to finish.

Take the time and go to the website, the article is short, concise, and should spark critical thought into one's own actions and abilities. In fact, you should continue to explore this topic. The more you read about it, the better your understanding of leadership currency will become.

Leadership currency facts:

- It is a finite resource.
- It is built up slowly but can be depleted quickly.
- Focusing on erroneous matters, or details, that are not directly linked to your mission, the Commander's intent, and the long-term success/overall wellness of the team will quickly deplete your stored currency.

Choosing the hard right, over the easy wrong. Your ability to maintain discipline whether it is your own, or the team's, will set the tone for others to follow. We are not a spit-and-polish organization, focus on details commiserate with mission success... BUT, do not slack on SOPs, tasks, policies, or regulations that could place the safety of the men, or the mission success in jeopardy.

5-7 Enhancing Leadership topics.

Always be learning, in my day this meant cracking open a book, or discussing topics with my peers and SGMs. Although this still applies, today this task is made easier with **Podcasts**. These are an incredible source of information for our profession, and currently, there is a plethora of Podcasters to choose from. Many SF and SOF veterans have jumped on this band wagon to ensure that all the GWOT knowledge and experience is not lost to the next generation.

At the time this book was written, I had only been listening to Podcasts for about a year or so. If there is a book re-write in my future, I'll most likely suggest many more channels, and episodes.

For our specific profession and relevant to this chapter, here are a few channels I know of:

- The Pinelander
  - Episode 15, "Tactical Leadership", Paul reviews his acronym BLEMISH, provides us some insights not captured in his book.
  - Episode 37, "Leadership Pitfalls", covers many of the topics I put myself through.
- Pineland Underground
  - Episode 40, "Leadership in the shadows" is 100% about leadership, and worth your time.
- Team VTAC, SGM (Ret.) Kyle Lamb
  - Episode 116, with CSM Dorsh (10th Group CSM at the time of this book) and his notes on leadership
- SOFCAST
  - I just learned about this platform. You can find it with an internet search, or on the SOCOM portal.

A quick internet search will reveal which platforms are required to listen to these podcasts.

Additionally, I have listened to Jocko Willink's podcast. He reviews a few of the books I listed (and many more) in the Team Sergeants library. It is a solid time saver because he reviews these books with a leadership lens, which for me, sparked how I interpreted the information covered, and how I wanted to apply it in my life. He also has interviewed numorous GBs from the MACVSOG era through to GWOT.

5-7.1 **For the Regiment**. Do not hide from the Battalion CSM. If you plan to stay on (for the right reasons) he is identifying his "star E8s" in the formation for future SGMs.

With that said, you need to dig deep and conduct a no shit self-assessment. Do you want to climb the ladder? Or, do you want to remain close to the tactical level, and continue to pass on your expertise? Or is it time to hang up the funny green hat and prepare for retirement?

Many of us see the Team Sergeant role as the pinnacle of our career, and for good reason. Like many Team Sergeants I was focused down and in, only concerned with my team. It wasn't until three years post team time that I realized the importance of solid guys pushing up into the higher ranks. Primary reason, to serve as a mentor for the next generation of Team Sergeants. Secondly, to be the vanguard when the good idea fairy strikes, and when they forget we are a bottom-up driven organization.

If you determine this is the path for you, engage with your Company SGM early. Tell him the following statement "I want to be an SGM." As silly as it sounds, this works. They now know you are interested and will help guide you in this journey.

I witnessed more than a few fellow E8s make the above statement, and within a few weeks (after his Team Sergeant time) he was packing up, and headed off to 1SG position somewhere in USASOC. Point being if you are a solid candidate for SGM, you will receive help. Another thing to think about, there are only four or five 1SG positions located in your Group. If you want to keep moving up, you may need to leave the comfort of Group, and jump on any 1SG position available. There are a few other positions that are regarded as "King makers," such as Battalion / Group S3 NCOIC, and the CMF 18 HRC posting.

5-7.2 **Last words on networking**. Do not be the total gray man. You need to network with more than just your Company SGM and fellow Team Sergeants. Build relationships with:

- BSC/HSC/HHC/FSC 1SG
- Support Company section NCOICs
- Company, Battalion, and Group Operations Sergeants

- Staff and section officers
    - Staff Officers: S3, XO, S1, S4
- Expand your network to include the other Battalion's staff NCOs and officers if possible

Mentor your guys, take them along when you conduct face to face meets, and show them how to network. Lastly, encourage your Captain to build his network, including all the staff officers within Group, not just your Battalion.

Other places to develop friends - any office or workspace on the Group compound that has a butt in a chair. It is good to know all the available resources, you never know what kind of support you'll need to get the mission done. Use those admin weeks to go explore the compound, introduce yourself, ask questions, and take notes.

5-7.3 **Going to the Doc is not a sign of weakness**. One area that was overlooked during my time was going to the doc. Push the guys to start cataloging their issues now.

An easy button for you and the boys to capture med issues is the Periodic Health Assessment (PHA). Every year each SM is required to complete a PHA, although seen as a waste of time, it is a great place to input injuries obtained during that year. When you sit down with your provider – which is mandatory to close out the PHA - you can establish a plan of action for the issue(s) you listed on the form.

A sweet go to is the THOR3 Docs, they record the session and upload it into your medical records. THOR3's primary purpose is to get you back into the fight. I have not experienced, or heard of these Docs placing guys on profile, limiting their usability… unless absolutely necessary.

Or use the standard route and make an appointment with your provider. Ask for follow on care, they will push for tests, x-rays, etc. and establish a care plan. Again, I have not heard of our PAs or Docs placing anyone on a profile that limits us for teamwork… unless you are that broken.

Either option is important for ETS, a MED Board, or retirement.

When the time comes to get out, the PHA input, Doctor/PT/OT appointments provide the SM:

- Proof the injury exists, and the required treatment
- Continued treatment demonstrates how the injury still bothers you, and a history of treatment is recorded
- Continued treatment potentially equals a higher VA compensation (rightfully earned, not some BS a supply SPC who never left the wire and gets a 100% VA rating for PTSD…)

If you are close to, or thinking it is time for retirement. Do not forget about yourself, hit up the Battalion PA, get seen for all the bumps and bruises you have ignored throughout your career. If you are married, ask your better half to recall all the complaints you have thrown out there and forgotten about.

5-7.4 A smooth transition process starts 18-24 months before your final day in the Army. Around the 18-month mark, you should be filling out your DD 2807.[5] At the same time, you should be determining your replacement plan for the SGLV. Every family situation is different, some want life insurance and SBP, some want more control and will bypass life insurance and rely on investments. Regardless, if life insurance is part of your financial plan, get it set up BEFORE you get tied down to a CPAP machine, diagnosed with Anxiety, or any other un-insurable ailments.

Don't just take my word for it, reach out to the JANUS program run by The Donovan & Bank Non-profit. They have a fantastic program and will travel to your Group. Visit donovanbank.org to learn more.

Lastly, I would like to mention Dan Rayburn. He is a friend of the Regiment and only helps Green Berets on their way out of the Army (ETS, Retirement, Med board). Reach out via mail@danrayburn.com.

---

5 PRINCIPAL PURPOSE(S): The primary collection of this information is from individuals seeking to join the Armed Forces. The information collected on this form is used to assist DoD physicians in making determinations as to acceptability of applicants for military service and verifies disqualifying medical condition(s) noted on the prescreening form (DD 2807-2). An additional collection of information using this form occurs when a Medical Evaluation Board is convened to determine the medical fitness of a current member and if separation is warranted. Completed forms are covered by recruiting, medical evaluation board, and official military personnel file SORNs maintained by each of the Services.

6

A Team Sergeant's Library

Leadership Principle number 2: Know yourself and seek self-improvement

6-1 Look to the past. Luckily, many have come before us and faced similar situations. Even better, they have captured their experiences in the written word. Fact, the more you read about your job, the more proficient you will become at it. Unfortunately, the perfect leadership book does not exist, and there is not a flow chart for every scenario. The only way to prepare yourself is continuous study in the art of leadership. Don't be like me, consuming leadership books, Podcasts, and reviewing historical references after your team time.

In this chapter I have listed the books that provided, in my opinion, the best information for our job as a Team Sergeant. In addition to the books listed later in the chapter, start asking those leaders you admire what they have read. No reason to dive headfirst into a random "Commanders recommended reading list." Those leaders you respect will tell you which books they have read, then tell you WHY those volumes are important.

Reading a book cover to cover is not everyone's forte. I strongly recommend you, and your guys, invest in audible books. Use the time required to commute to work, long runs, road marches, or yardwork to improve your knowledge base.

6-2 Team SOP book. The first book I will cover in this chapter is one of the most important books you should have available on your ODA... The SOP Book. If your team does not have one, take the time to create one. You do not have to develop this product on your own. You have smart guys on the team, incorporate them in the development and updating process. Including the guys does a few things, one you get a snapshot of their understanding of each topic in the SOP book. Two, they will have a feeling of ownership on the team. As with any manual in the Army, it needs to be reviewed and updated often.

You have the final say for all SOPs but remember to let the boys experiment with new ideas. They may have been on a squared away team prior or attended a course that provided a technique that could simplify some complex tasks. Don't get stuck in the Dogma of "this is the way I have always done it."

If you are starting from scratch remember you are not the first guy to create one of these books. Talk to your peers and dig in the I Drive. Find a template to get started. Once it is completed, build your checklists[1] (PCCs) with each section of the SOP book.

---

[1] See Kyle Lamb's book *Leadership in the Shadows*, Chapter 36, for a detailed breakdown on checklists.

Lastly, this should be the first reading assignment for new guys showing up to the team.

Below MSG Longoria and I have listed (in no particular order) a few categories included in our SOP books.

- Duties and responsibilities by MOS. (see Annex1, page 329 and 330). This MOS responsibilities breakdown clearly identifies who is responsible for what, during the planning cycle. Whether you are using old school TLPs, building a CONOP, or MDMP.
- Formations, and Order of Movement (FOOM) diagrams.
    - Both dismounted and mounted patrolling in various environments (urban, rural, forested, dessert, etc.).
        - For mounted, plan for two, three, and four ODA vehicle convoys, think about PF vehicles. Include all known platforms. You never know when you will be operating out of civilian / low vis vehicles. Obviously as new tech is introduced these will change or add additional pages.
    - Actions at danger areas
    - React to contact
    - React to an ambush (near/far)
    - React to IDF
    - Down driver drill
    - Down gunner drills

  - Flat tire drills to include under NODs
  - vehicle break out drills
- Specialty infiltration methods/skill-set SOPs. Don't re-write something that exists. For example, on a mountain team, there is a manual produced by the schoolhouse that covers everything from how to wear your clothing system, through how to set up and employ complex rope systems.
  What you need to identify is: which guy/MOS/skill level is responsible for what in that specific complex rope system set up, and employment. Or, how the detachment is broken down into rope teams, the shared equipment they all carry to create the system. Who is going to be on security, etc.
- Hand and arm signals. You do not have to recreate the wheel here. If you cannot find anything already built on the portal, open internet, I Drive, etc., reach out to the Master trainers in B Co, 1/1 SWTG(A).
- Military Vehicle[2] set up and load outs. This can be as detailed as you need it to be. Try to establish a standard that covers a majority of GAF mission sets. Stay flexible, as we know requirements will change. There are many details to consider, everyone can chip in with their expertise. Just focus on one aspect at a time, and soon the whole

[2] You may not have all the answers, so lean on your senior guys, or other Team Sergeants. Always conduct a share drive search.

picture will come together. Below I listed out a few items that we had to create solutions for in the past.

- Ammunition. Consider details like ammo quantities per weapon system and storage. Does the load out make sense for the gunner who will need access to cans for reloading under the worst conditions?

- Communications gear. Can the vehicle support your communication PACE plan? Do you need the tech shop, and the metal fabricators to help your 18E configure the trucks?

- Blow out bags. Where do you keep them? What is the packing list? Are there generic bags always on board, or is each guy responsible for his own bag?

- Driver, TC, gunner, etc., responsibilities pre-mission, during, and post op. Describe which equipment is he responsible to fire up, shut down, and how to do it.

- Tow or push a dead truck out of the kill zone... This one has swung many ways. I have seen a few methods, and had a heated discussion or two regarding this topic. The important part is this, do your homework for your AO, figure out the least complex way to get out of the kill zone quickly, and rehearse. Rehearse until each guy can do every position

  - Convoy net SOPs. Right turn open from truck 1, right turn closed from truck 4. Truck 1 check point 3 open, truck 4 check point 3 closed, etc.
  - Where are key leaders in the convoy?
  - How do you divide up the MOSs?
  - Have a Task Org for assigned US uplift (drivers, gunners, and some dismounts).
  - Same for zero uplift personnel. Yes, before the days of uplift and contracted security, the ODA did it all.

- ISOFAC set up. Does your Group maintain a facility for this task? If not, does your team have the essential equipment available to them? If so, where does this stuff reside? Your 18C should have this information readily available in his master property book. Outside of the furniture requirements, be sure to have the obligatory SF doctrine (digital or physical), and planning product templates.
- Emergency Deployment Readiness Exercise (EDRE) procedures. If I could do it again, I would set up my team room like a ready room. Create a SOP of minimum equipment to have on hand at all times. Minus obvious times/events such as schools, or using kit on the weekends for personal range ops.

    - Once set up, I would incorporate EDREs as much as possible. Nothing hones your SOPs and checklists like a tight time hack! Additionally, EDREs test your phone recall plan, and the team's response time.
- Kit, so many ways this can go down. Each mission set is different, take the time to develop YOUR minimum equipment requirements to conduct specific training and/or mission tasks. I fully supported shooter preference, but their kit had to have my required items/basic loads. Make sure your seniors are working with the new dudes. Provide the team range time to work through kit set ups, this is where many adjustments will happen.

6-3 Team Library. Manuals to have on hand in the team room or know where to find them online. Look on USASOC home page for the ARSOF Doctrine tab.

> **NOTE: These manuals are updated frequently. By the time this book is published, many of these manuals will have updated copies.**

*I referenced the 24 JAN 2019 version.

**Be familiar with = read the Summary, applicable paragraphs, and the initial paragraph in each chapter.

1) *1st SFC(A) Regulation 350-1
2) **FM 3-18 SF Operations

3) ATP 3-18.1 SF UW (CH 1-6 and Appendixes A-Z/two books)
4) ATP 3-18.4 SF Special Reconnaissance
5) ATP 3-18.10 SF Air Operations
6) ATP 3-18.14 SF Vehicle Mounted Operations
7) **TC 18-01.1 UW Mission Planning Guide ODA-level
8) **TC 18-02 SF Advisor Guide
9) **TC 18-05 SF FID and Support to COIN
10) **TC 18-18 SF Handbook
11) TC 18-20 SF Site Exploitation
12) **TC 18-22 SF Training Strategy
13) TC 18-32 SF Sniper
14) **GTA 31-01-003 Detachment Mission Planning Guide
15) GTA 31-02-001 SF Air Operations
16) GTA 31-02-002 Air Tasking Order and SPINS
17) **STP 31-18-SM-TG SM-TG SF Common Skills, levels 3/4, (correct you will not find a version newer than early 2000s)
18) **STP 31-18B34-SM-TG
    STP 31-18C34-SM-TG
    and FM 5-34 Engineer Field Data
    STP 31-18D34-SM-TG
    STP 31-18E34-SM-TG
    STP 31-18F34-SM-TG
19) **USASOC 350-2
20) AR and DA-PAM 670-1, yes, you need to know how the uniform stuff works. Most guys do not care, and you will have to inspect everyone before anything

important. Good rule of thumb, every NCO should have their dress uniform dry cleaned, and set up on a hangar. Boots can be shined quickly. NCOPD, have the guys look up how to correct the deficiencies when you find them.

21) **FM 1-02.2 Military symbols
22) **FM 1-02.1 Operational Terms
23) ATP 3-05.1 UW at the JSOTF lvl
24)** ATP 2-01.3 IPB
25) **TC 18-05 SF FID and support to COIN
26)**TC 18-01.1 UW Mission Planning Guide for the SFOD-A level
27) FM 6-22, Appendix B (Counseling)
28) TC 21-76, Ranger Handbook

Note: Again, doctrine titles, numbers, topics will change over time. Always do your research to identify the most current publications. Also, refer to DA PAM 600-25 for the most current list of required manuals for each MOS to know.

6-4 My recommended list. Books and manuals I wish I had read prior to my Team Sergeant time:

1) 1st SFC(A) Regulation 350-1

2) Your specialty infill regulation, example: USASOC 350-12 for Mountain Teams.

3) DA PAM 600-25 lays out the 18 Career Management Field (CMF), it is the basis for the Professional

Development Model that CSMs hand out during NCOPD events.

4) TC 7-22.7 The Non-Commissioned Officer Guide, Jan 2020, not a cover to cover read, but filled with all that big Army stuff we need to know.

5) *Leadership in the Shadows* by Kyle Lamb.

6)*US Army Small Unit Tactics Handbook* and *Tactical Leadership* by Paul LeFavor. I bought the SUT book after my team time, specifically for the leadership and precise history of SOF chapters, ended up going cover to cover and hating that I did not read it prior... reason being, it had not been published yet.

7) *Always Endeavor, a Developmental Guide for In Extremis Leaders* by Colin Greata. As I mentioned in Chapter 2, his book is fantastic for establishing your initial counseling.

8) *About Face* by COL (ret) David Hackworth. The COL puts a lot of emphasis on spit and polish but, his combat experiences, common-sense approach to leadership, training for war, and going to war is invaluable. It is a monster book, or ... listen to Jocko Podcasts #2, 249, and 250.

9) *From OSS to Green Berets* by COL Aaron Bank. You need to know where you come from.

10) *Afghanistan and the Troubled Future of UW* by Rothstein.

11) *Fire in the Jungle* by Schmidt. A fantastic account of the UW campaign in the Philippines during WWII.

12) *Tactical Checklists* by Bronston Clough This book would have been great to have when I was building my SOP book.

13) *Once an Eagle* by Anton Myrer. Although based on a fictional character, and an officer, the leadership qualities he embodies hold true to this day.

14) *Call Sign Chaos* by Jim Mattis. I wanted to read his account of what happened early on in Afghanistan, due to some controversy with the ODAs on the ground. Surprisingly, I came away with some common-sense leadership tips I added to my tool kit.

15) *The Mission, the Men, and Me* by Pete Blabber.

16) *Leadership, the Warriors Art* by Kolenda.

17) *Conversations with Dick Winters* by Dick Winters, or listen to Jocko Podcast #75.

18) *The Warrior Ethos* and *Gates of Fire* by Steven Pressfield.

19) Any books published by GBs, or written about GBs in action:

a. *Not a Good Day to Die. The untold story of OPERATION Anaconda* by Sean Naylor.

b. *Gentlemen Bastards, on the ground in Afghanistan with America's Elite Special Forces* by Kevin Maurer.

c. *Code Name Copperhead: My true-life exploits as a Special Forces Soldier* by Avrum M. Fine and Joe R. Garner.

d. *Horse Soldiers* by Doug Stanton.

e. *Swords of Lightning: Green Beret Horse Soldiers and America's Response to 9/11* by Mark Nutsch, Bob Pennington, and Jim DeFelice.

f. *The Green Berets* by Robin Moore.

g. *The Only Thing Worth Dying For* by Eric Blehm.

h. *Masters of Chaos. The Secret History of the Special Forces* by Linda Robinson

i. *Legend* by Eric Blehm. The recounting of Roy Benavidez's actions that earned him the Medal of honor.

20) Enhancing books:

a. *On Killing* and *On Combat* by LTC Grossman.

b. *Leaders Eat Last* and *Start with Why* by Simon Senik. If you want to save time, he has a butt load of YouTube videos.

c. *Tribe* and *War* by Sebastian Junger.

d. *Noise* and *Brief* by Joseph McCormack. I had the opportunity to attend a one-day lab on the principles, and application of the skills in his books. As a Team Sergeant you will be expected to brief the Battalion Commander on a regular basis. We as NCOs, have a tendency to go super deep into the weeds. Although that is our job, when you are briefing the higher ups there is a certain way to do it. With that said, these books will help you refine your approach and delivery.

Additionally, these books are a good tool for you while working with the Captain on selling the team. I don't think I can oversell the concept of "reverse mentorship."

e. *Training For the New Alpinism* by Steve House and Scot Johnston. If you are running a Mountain ODA, this is a must read when building your PT program.

6-5 To close this chapter out, the bottom line is you have to continue to educate yourself. The NCOES system has gotten better over the years, but you are responsible for your own path. Whether you read the suggested books in this chapter, or get after it with other selections, continue to improve your understanding of military history, leadership, and all things related.

7

Parting Shots

7-1 The following is a medley of **pro tips** in no particular order of importance. I could not find a home for these in the previous chapters. Most of the info is mission enhancing, but some of it will keep you off the bosses' carpet in your best uniform.

7-2 DIRLAUTH. Direct Liaison Authorized is a thing to know and understand. Every Group's chain of command, and their staff handle DIRLAUTH differently. Be sure to ask if you have it before calling up and out. My Group was super picky during my tenure. It was almost a mortal sin when a Team tried to work directly with staff at Battalion and Group.

On a similar note, be cautious when you receive a call from two, three and higher levels up. Your organization may not care, but mine did. Let someone know when a USASOC Staff member calls down to your Team room.

7-3 Email etiquette. Emails, the official correspondence method of today's Army. You will receive and send plenty of these... so will your men. Here are a few things for you and the team to do and avoid.

- Ensure everyone has a signature block set up to auto populate on every email they generate.
    - Rank, Name
    - Unit
    - Email accounts
    - Work and cell phone
- Emails are a fantastic way to pass information to one, or many people. When it comes to forwarding, do it with a purpose. Provide a quick summary of the info in the email. This quick writeup will give some context to those you are forwarding to.
- If you send an important message to anyone on, or outside of the team, follow up with a phone call.
    - If the task is time sensitive, provide that person everything they need to know about the task, especially a no later than time.
    - Much is lost in emails (and texts), discussing your message face to face/over the phone will help fill in those critical gaps.
- Up and out communications. If your guys are sending emails outside the team, it is a direct reflection of your ODA.
    - You do not have time to scrub every email they send, do not try (directly tied to: give trust, to build trust). You and the Captain need to establish a team SOP on what your team's correspondence should consist of... and what is worthy of being cc'd on.

7-4 Free money. Always have an end of year funds wish list built. When the fiscal year comes to a close, the unit S4 normally will announce that they need to exhaust remaining funds. This request always comes down when you are busy, and do not have time to develop a proper product.

Compile all the mission enhancing items that did not make the cut on previous requests. Also known as the super "important" Gucci gear the guys wanted for trip X. Templates of approved requests should be on the shared drive or portal. If not, energize the B-TEAM to obtain a copy. This is a perfect task for your 18C.

7-5 Team funds are a must have. This cash will cover down on all the cool crap you want to do like:

- Team patches
- Shirts, hoodies, hats, etc.
- Coins
- Going away gifts
- Gifts for civilian organizations/land managers
- Food for the guys on those late-night planning sessions

If someone does not want to participate, no problem, a team fund is voluntary... of course that guy may or may not get a PCS/ETS/ Retirement gift.

I used my JCET FOO/PA, normally an 18C, for managing the team fund.

7-6 Team property. Your 18Cs should have a bad ass property book, and a functioning stay/go list. The property book should contain enough information so anyone can find the 10% monthly inventory items. At a minimum each item on the property book should have a location written next to it. Preferably, a monster property book is created, where each item has a picture, along with the storage location listed.

7.6.1 On the topic of 10% inventories, my Company was famous for the no heads up inventory. I would receive a call from the B-TEAM supply Sergeant asking to see X, Y, and Z... just as I arrive in Vail, CO. A mere 3-hour drive from FT Carson.

A possible fix is to be proactive. Lean on the Company supply sergeant to get those dates in advance. Coordinate with him or the Commander to knock inventories out prior to training events. Worst case, you leave someone behind and he catches up to the Team at the range.

7-6.2 Always obtain a turn-in document, even if it is a hand jammed DA Form 2062. Never discard property turn-in documents, keep copies of these in the property book forever.

7-7 Knowledge management. Set up your shared drive and/or portal page so it is easy to navigate. If you find a gold mine (another 18Z's file) set up a shortcut to that folder. My Group's IT shop was notorious for "cleaning up" the I drive without warning. If you cut and paste all the good stuff, your folders will be deemed too large, and will be trimmed down.

Another thought on knowledge management, do not be, or let anyone on the team become the single point of failure. Saving team documents to your desktop is a no go. With today's tech, you should have anywhere access with an issued laptop and VPN, no excuses.

7-8 Ethics within Special Forces. In the past few years, there have been speeches, policy changes, articles, and academic classes created around this issue. The top three have an obligation to do your homework here. Do what you can to include this topic in your FMPs. If that idea seems canned, see what is available (online or resident classroom-JSOU) for you and/or the team to attend. You can do one hundred things right, but one ethics issue is all it takes to dismantle a team.

Ethics on the Detachment starts with you. Every time you are faced with an ethical decision, the action(s) you take will set the tone for the ODA to follow. Every time one of (or a group of) team members step off the moral/ethical azimuth you must step up and address it. You will be measured against what you let slide on the ODA.

The team is not the only ones affected. The unit, and even the Regiment will take a hit depending on what transpired. When events like this happen, all the negative light is cast upon us. This will force Commanders at all levels to take control of us at the tactical level.

7-8.1 *Special Warfare* magazine, January-March 2022, volume 35, issue 1 has three articles that are worth your time regarding this matter.

- Developing SOF Moral Reasoning
- Ethics is Leader Business
- All Training is Ethics Training

7-9 Outside help. Create a list of resources /people/offices you can call for specific situations. Lean on the SGM, and if you have a good relationship with the 1SG, for this information. You do not have to solve all of the world's problems. Once you know who can help you or your men, go visit them and introduce yourself. People respond better when they have a face to match to a name and voice on the phone.

7-10 Personal responsibility/time management. Build and maintain an "I love me book." This should not be new information for you, but important for your guys if they have not built one. S1 is going to lose everything you (and your guys) give them at least once. An I love me book is a steadfast way to do their job for them... I mean provide S1 the documents they require.

After the third or fourth time the B-TEAM requested the same document for Battalion, I started a folder on the I Drive that housed every digital document I sent. Then I shared the link with the B-TEAM Ops sergeant, so he could retrieve what he needed... again.

7-11 CAO/CNO. Casualty Assistance Officer/Casualty Notification Officer. Although you should not be an active one as a Team Sergeant, you should get the training. This will open your eyes to the importance of the DD93 and SGLV forms, and better help you to counsel your guys on how to properly fill them out. Losing a teammate is hard enough, do not let it be worse because he never updated his paperwork… and now his ex-wife is getting everything.

7-12 Leave forms. Every Battalion's S1 does this a little different. Have the 18D interface with the Company, or Battalion to get the skinny on the unit's SOP. He should also be in charge of instructing the ODA how to fill out the DA Form 31. Additionally, he should be responsible for scrubbing, submitting, and tracking until approved. As with anything, be sure to spot check a few before starting the signature process.

When leave is submitted outside the ODA's planned block times, review your training calendar for the requested days. If that SM is critical for a training concept (i.e., your only 18D), be prepared to coordinate a replacement from down the hall.

Common issues to look for are:

- No LES attached to email, or preferably attached to the DA-31 itself
- Wrong number of leave days available used from LES on the DA-31
- No digital signature of the SM

- Wrong type of leave requested
- Wrong Origin address

7-12.1 While I was finishing this book IPPS-A came online and I am unsure if my leave form suggestions are still relevant. Regardless, you and the 18D should become intimately familiar with the new system.

7-13 Feasibility Assessment. Every AOR has a hot spot or two, this can trigger a short suspense deployment. Don't get caught with your pants down, engage with the AOB and Battalion staff for a product template.

Example: My ODA was in the collective training phase of our LRTC, our mission was a JCET at the end of that year. In fact, half of the Company was on the same timeline. At one point between August and October of 2011, in a north African country, a well-known dictator had lost power, and was on the run. I can only speak for my Company, but three ODAs were tasked with preparing a feasibility assessment for a potential short suspense deployment. I am not sure what your Group would want to see, but here are a few things we prepared:

- Individual/advance skills chart, that included Language capabilities, college degree info, Combat experience, Team time, etc.
- Loss and gains roster, to include SMs projected to attend schools
- Current and projected T, P, U ratings for collective tasks

- Critical equipment shortages

Lesson learned, have a template filled out. Update it whenever the ODA completes a phase of training - add this to your battle rhythm checklist. Check in with Battalion a few times in the year to see if changes are required. A new Battalion Commander can trigger these changes or, if an event has happened in your AOR, and another team has completed this process.

7-14 Common sense. Take the trash out before weekends, and long trips... fruit flies are terrible.

8

Time to Take the Reins

8-1 Hand over checklist. In Chapter 3 I touched on the topic of not knowing the complete picture when you step into your new role. Whether you are replacing one of the top Team Sergeants in the Battalion, inheriting a team after an implosion, or anywhere in between, you need to get your bearings quickly. Heads up, if the outgoing guy was nuked in place, then you may not get the chance to interact with him at all.

However, your handover situation plays out, below I have provided a basic checklist to focus on during your right seat / left seat ride.

1. Before you sit down with the outgoing guy secure a copy of and understand the Commander's Annual Training Guidance. Get your eyes on the Battalion's LRTC and connect how the Annual Training Guidance ties into the calendar. Begin to ask important questions such as: Where (geographically) is your team going? What is the mission? What if your team does not have a mission? Ask if this is a re-build year for your team. You might be able to do this as soon as you have pinpoint orders, while in-processing at Battalion.

2. Sit down with the Company SGM. He should be able to answer the following questions:

- What are the dynamics on the team?
- Was the Team Sergeant a strong leader, or did the 18A, 180A and/or a senior E7 run the show?
- Has the team suffered any recent losses (combat, training, or otherwise)?
- Is the manning at full strength? If not, when can we expect the next round of gains coming into the Company?
- Projected losses to PCS/ETS/Retirement?

3. Once you get down to the team, whether you sit down with the outgoing guy, or are engaging with the 18A/180A, or a squared away E7, ask to see the team's LRTC. Does the glide path make sense for the mission they have been assigned? What training phase is the ODA currently in?

4. Review the hour-by-hour training calendars, there should be six-weeks or more prepared. What is on deck in the upcoming weeks? Ask if the team is ready for these events. Are there any outstanding resources, or last-minute coordination's to be made?

Is there a checklist covering these items, and individuals assigned? If nothing resembling a checklist exists, or the team seems disheveled, dig up and review the training concept for the upcoming event. Then prioritize, execute, and supervise.

5. Get the manning breakdown from the outgoing guy. Same questions as with the SGM, what are the projected losses to PCS/ETS/Retirement? By asking the SGM, and the outgoing Team Sergeant, you can ascertain if the Company and the ODA are synced.

6. What is the NCOER closeout plan? The outgoing guy should have a list of his upcoming NCOERs (annuals, CORs, Extended annuals, etc.). If the SGM approves, you might inherit a few annual NCOERs if they are due 90 (or more) days past your start date.

If this is the case, ask the outgoing guy to start a fillable PDF 2166-9-2, with admin data, and at least one bullet per section. This will serve as the letter of continuity. Lesson learned, don't let the guy depart without NCOERs being done and submitted to the review chain.

7. Same thing with awards. The guys will work hard for you, just like they did for him. The right thing is to get his input before he takes off.

8. Team personality/Culture? Is this the CrossFit team, or the close to burnout team (always gone to training) in the Company? Is the team super chill, or are they former Ranger Battalion heavy/eat their own?

9. Who on the team has the strong personalities, and who are the followers? Is there a possibility for a shadow government?

10. Is anyone going through a divorce? Flagged? Legal issues? Money/debt issues? What is the status of these individuals? Do they have, or know of all the resources available to work through their situation?

11. What is the overall fitness of the ODA? This initial snapshot can be ascertained by reviewing the previous six-week calendars (scheduled team PT). Are the ACFT Scores/team average posted? Of course, the truth will come out during your initial assessment period and when you conduct the first ACFT.

12. Ask to see the Team's advanced skills tracker, does the team have the minimum skills and qualifications in accordance with 350-1? If not, and the ODA is in the Individual skills phase, do guys have those required schools and/or courses lined up?

If the Team is deeper into the LRTC, is there a plan to get people trained up/re-certified? Or have we identified a possible loaner body from another ODA?

13. What kind of leaders are the Captain and/or 180? Do they have a good reputation with the Company and Battalion?

14. What is the ODA's battle rhythm? This one is important, no reason to shake things up in the first few weeks before you have completed your assessment period. If there is a major safety concern absolutely intervene.

If no safety issues exist, let the team do its thing. How many times have you had a new leader walk in, and they start making changes without knowing anything about the team? Never forget that frustration and treat your new team how you wanted to be treated.

8-1.1 As you will find out, the above check list is not complete. It is merely a tool to get to know your new operational environment. You will need to have a similar conversation with the Captain, 180, and senior guys on the team.

8-1.2 Conversely, when you are preparing for your exit, revisit this list. Prepare a handover plan, drop "packet" for your replacement.

8-2 The 12 Principles in action. During the handover, and early on in the assessment period, pair the above check list with the Twelve Principles of Leadership, CSM Dorsh's Notes on Leadership, and COL Johns' Leadership principles in Appendix II. By doing this, a clear picture of the ODA will appear.

Knowing if these principles were used by the previous Team Sergeant is important information. If they were, then you may be the one adapting to what already works for the guys. If not, then you will need to have a method ready to implement soon after taking over.

Below I have paired directly or tied to the spirit of the 12 Principles with the majority of the Handover Checklist. This pairing is based on my experience, you may find a different situation on your ODA.

Principle 5. Is there a current LRTC and an updated six-week calendar posted in the team room? If so, then principle number 5. Keep your men informed, is a staple on the ODA.

Principles 1 & 7: What is on the calendar? Is it primarily focused on METs needed for the upcoming mission, or is it filled with erroneous cool guy stuff? Does it appear that the team's leadership even read the Annual Training Guidance? Ditto for team's advanced skills tracker. This observation will clue you into the team's mindset on principles number 1. Be technically and tactically proficient, and 7. Employ your team in accordance with its capabilities.

Principles 3 & 10: If the outgoing guy spends a majority of the handover blaming everyone else for the ODA's shortcomings, then you can bet the men (maybe not all of them) will do the same. Take note, you may need to apply energy to establish an environment where everyone focuses on number 3. Seeks responsibility, and takes responsibility for their actions. And for you, number 10. Set the example. Sitting down with the SGM prior should prepare you for the above possibility. Again, knowing your operational environment isn't just for deployments.

Principles 8 & 9: Ensure each task is understood, supervised, and accomplished, and number 9. Develop a sense of responsibility in your subordinates. Is there a MOS task board, or anything resembling a tracker for upcoming concepts? Are the tasks assigned by name? Hopefully the answer is yes, otherwise the team was possibly run by a micromanager.

Principle 11: Build a team. It will not take long to see if the ODA functions well together. Everything from garrison to range work will expose any infighting that exists. No need for me to ramble on here. Use the tools listed in Chapters 1, 2, and 3 to start building your team.

Principle 2: Did the former boss encourage Know yourself and seek self-improvement? If there was something in place do what you can to keep that going. If not, encourage the guys to read, listen to podcasts, and attend seminars held by members within our community. Push them towards anything to broaden their understanding of the profession.

An easy target for this leadership principle, assign a Podcast to the team. Give them a week to listen to it, then sit down at lunch and discuss it. Show up prepared for the talk, have questions that get them involved. For example, if you instructed them to listen to Team VTAC, episode 116 with CSM Dorsh, I would ask: What does – "Competition breeds success. Man in the arena. Earn it, tomorrow is not a right" - mean to you? Or, which leadership note that was talked about struck home for you the most? Or was there a leadership note that you think is being missed on our team?

Does the team have a physical, or digital library? If not, do they know where to find all of our manuals? Army Publishing Directorate does not carry all of our stuff. Do they know which manuals they should have readily available? DA PAM 600-25 lists out every manual each MOS should have access to, make sure your guys know about this manual.

As for the guys reading a book cover to cover, persistent recommendation does not always work. I found by telling the stories, and the lessons I pulled from a book sparked conversation. These conversations actually led to some guys becoming excited to read it for themselves.

The last three principles are difficult to assess, but not impossible. As you begin to implement your version of the following principles, there will be clues in the men's reactions if these were practiced or not.

Principle 4: If Know your men and look out for their well-being was a staple on the ODA, the outgoing guy will have the NCOER handover lined up. Annuals, CORs, CTR, Letters of Continuity, etc. will be prepared. If he was super squared away, then letters of continuity for awards will be done as well.

Along with evaluations, there should be a folder on the shared drive, or a counseling section within the leader's book. You shouldn't get to see the outgoing Team Sergeant's counseling forms on the guys, but if the team guys act surprised that you are counseling them (or utilizing NCOER Support forms), then you know it was not done.

Just as important as the above items, if not more so, the outgoing guy should have answers to handover checklist items 8, 9, 10, and 13.

Principles 6 & 12: Create a positive climate, as time rolls by, and you observe the men's day to day actions, you will learn if this principle was followed. If the culture and overall team vibe is good, then slowly begin to interject your version of leadership into the team. No reason to fix something that isn't broken, continue to add fuel, and keep that machine supplied with oil.

If not, you may need to jump in quickly to establish control and order. Below, I broke down Paul Lefavor's proposed 12th principle into separate sections (a./b./c./d.). For each section I included my understanding of them based on experience.

Before I conduct my deep dive, let this quote set the stage for this principle:

> "The role of a leader is not to come up with all the great ideas. The role of the leader is to create an environment in which great ideas can happen."
> - Simon Sinek[1]

[1] Simon Sinek, Start with Why: How Great Leaders Inspire Everyone to Take Action.

a. Good leaders understand that they set the moral and psychological thermostat in both battle and the workspace. This principle is relative to what is referred to in the military as 'presence.'...

It is possible for the ODA to adopt many of your qualities, and overall attitude. This is important to understand because if you have a crap attitude when it comes to the mission, so will the boys.

You will not agree with everything that comes down from higher, and not every mission we do is sexy. When those orders, events, missions, policy changes, etc. come down the pipe, go fight the fight in the Commander's/SGM's office. State your case, tell them how you really feel, explain how it negatively effects your team, and/or your mission. This generally works best if you provide solutions for the boss to consider.

After you have said your piece though, when it becomes an official order, own it. Go back to the Team Room and be the adult in the room. If you are able to give a WHY to the boys, sweet. If not, sometimes we all just have to accept the fact that we are in the Army.

On the other side of the above topic, sometimes the boss is right. Here is what Kyle Lamb has to say:

> But what if the man is right? Standing up for a group of your peers is often easier than taking the unpopular position by standing up in the

> opposition to those same peers. If your superiors have made the correct decision, you must do the same thing and stand with them.[2]

With all that said, if the last guy did not adhere to the above mindset, the men may be resistant to your approach. Be prepared to stand your ground.

b. This principle reminds us that good leaders galvanize the strengths of others. Thus, rather than seeing strengths in others a threat to our leadership, good leaders envision how to implement those strengths and motivate their subordinates to use them...

Use the strengths of your guys to best accomplish your mission. Also, use those strengths to balance out the other guys, and if you have to, your weaknesses. Sometimes you will see a talent someone has, but they do not realize it. You might have to push them in that direction. Basically, not everyone is a shooter, some of us are geared for intel activities, or something in the tech world.

c. Moreover, control and order are immediately reestablished by presence.[3]...

This comes into play if you are taking over a team that one, or all of the leadership was nuked in place. You still need to conduct your assessment, and selectively initiate

---

[2] Kyle Lamb, Leadership in the Shadows, chapter 13, page 61.
[3] Taylor, *Military Leadership*, 36.

changes, but you need to provide immediate guidance for that team.

This immediate guidance ties into principle number 6. Make sound and timely decisions. Before you take over, read and understand the Commander's Annual Training Guidance. If you understand the Commander's intent, then you can make decisions quickly.

Just like react to contact, provide the team a Distance, Direction, and Description! Give the team a short term, achievable goal to strive for. As small victories happen, the team begins to heal, and comes back together. Now you and the Captain can expand your left and right limits and push your end state out further.

d. A critical component of presence is composure, that is, a leader's command over himself.

When shit hits the fan, the team looks to you. Whether your Company leadership dropped an EDRE on your team at 1700 on a Wednesday night, or a small TIC becomes a complex attack, the team 100% relies on your composure.

Cool as a cucumber was the phrase I grew up with. I never fully understand what that meant, until I saw it in action. Back in the day, my first deployment after the Q course took me to Kosovo. I get it, of all the stories to tell in order to drive this point home, I chose this one - This trip was a welcome break for my teammates, who had just done back-to-back trips to OIF I and II.

The mission itself was not difficult. The trip provided ample time to execute team training, and for the guys to get back in the gym. Not a bad trip for me, the FNG, but I would be lying if I said I was not disappointed with my first deployment's location.

Our biggest challenge was combating boredom. One night, deep into a Madden football marathon on the PS2, a loud and thunderous boom shook the house. Being the newb, I was confused on what my next move was. I heard the team begin to hustle around, and that feeling of being wrong began to sink in.

I was leaving the TV room when my Team Sergeant made eye contact with me. He plainly stated, "grab your kit, blow out bag, and load them in your truck. Next, help the medic grab extra supplies." Within a few minutes, the team was outside conducting a truck side brief.

The leadership informed the team that shortly after the blast we received multiple calls from our interpreters and partner force officers who lived in the immediate area. They were all calling to explain that the local shopping center (think multi-story PX with an abundance of marble) had been hit with a VBIED. They reported that the blast had created a MASCAL situation. The rest of the brief was simple, just a 5Ws and a quick rock drill that got us out the gate.

Upon our arrival, my Team Sergeant gave simple instructions to the 18Bs to secure the scene and helped establish the perimeter. With security set, he then assisted the 18D with the injured personnel. As it turned out, the mayhem that was described, wound up being two guys with some cuts and bruises. One cut was pretty close to the jugular but was easily managed by our medic.

As the Team Sergeant and Medic did their thing, the Detachment Commander was reporting up and out. He too was calm and staying in his lane. This calmness kept me, the new kid, and everyone else focused on the tasks at hand. After what felt like a few hours, the Captain was able to coordinate transportation of the injured persons to Camp Bondsteel, and we handed off security of the scene to the host nation police.

Although this happened on a non-kinetic trip, the situation could have quickly spiraled out of control. Your ability to remain calm, cool, and collected will ensure the ODA will too.

8-3 Everyone's experience will be different during the handover and initial assessment. Review, and plan to use this checklist as a guide through this fast paced, and exciting time.

Closing

My final lesson learned... after your time in the seat is done, do not get stuck in your head and focus only on the negative. You have to acknowledge your wins along with your losses.

Every guy who has done this job can either relate with my failures or has tales of their own to share. I strongly believe we have an obligation to share our leadership struggles with those wanting to take the reins. On the other hand, we all have our positive accounts, and we are responsible to share those as well.

At the beginning of my journey in writing this book, I was 100% convinced that I failed almost everything I did as a Team Sergeant. In every professional development book I read (post team time), I found countless principles or actions I know I screwed up. This information pushed my writings to be hyper critical of my failures.

As the book began to take shape, I shared it with trusted peers, and Sergeants Major. They all had a common theme as they came back with suggestions... "why all the self-deprecating talk?" These shared observations opened my eyes to all the good that came during my time. They helped me concentrate on what I have learned, and how I could share my experiences.

I was fortunate to be assigned to B Co, 1/1 SWTG(A) prior to retirement. I was surrounded by successful post team time E8s from the whole Regiment, and they were invaluable in the development of this book.

Although the data and stories they provided opened my aperture from a 10th Group specific, to a Regimental lens, our information may not always be relevant to the future generations.

For this book to remain relevant I need those who have recently done the job to reach out. My experiences and TTPs will most likely become outdated and replaced. Send in your updated information, share a better story to highlight an important point, or, let me know if I totally missed a topic that is critical for the up-and-coming Team Sergeants.

If you have something to share reach out to the publisher.[4] They will ensure I receive your message, and we will take it from there.

It is my sincere hope this book serves you well, and that you feel confident in recommending it to those who are worthy of running an ODA in the future.

De Oppresso Liber

[4] See pages 345 of this book for contact information.

Appendix I
Special Forces Creed [1]

I am an American Special Forces soldier. A professional!

I will do all that my nation requires of me. I am a volunteer, knowing well the hazards of my profession.

I serve with the memory of those who have gone before me: Roger's Rangers, Francis Marion, Mosby's Rangers, the First Special Service Force and Ranger Battalions of World War II, The Airborne Ranger Companies of Korea. I pledge to uphold the honor and integrity of all I am - in all I do.

I am a professional soldier. I will teach and fight wherever my nation requires. I will strive always, to excel in every art and artifice of war.

I know that I will be called upon to perform tasks in isolation, far from familiar faces and voices, with the help and guidance of my God I will conquer my fears and succeed.

I will keep my mind and body clean, alert and strong, for this is my debt to those who depend upon me.

I will not fail those with whom I serve. I will not bring shame upon myself or the Forces.

I will maintain myself, my arms, and my equipment in an immaculate state as befits a Special Forces soldier.

My goal is to succeed in any mission - and live to succeed again.

I am a member of my nation's chosen soldiery. God grant that I may not be found wanting, that I will not fail this sacred trust.

De Oppresso Liber

---

[1] Special Forces Creed of 1961.

Non-Commissioned Officer (NCO) Creed

No one is more professional than I. I am a noncommissioned officer, a leader of Soldiers. As a noncommissioned officer, I realize that I am a member of a time-honored corps, which is known as "The Backbone of the Army." I am proud of the Corps of noncommissioned officers and will at all times conduct myself so as to bring credit upon the Corps, the military service and my country regardless of the situation in which I find myself. I will not use my grade or position to attain pleasure, profit, or personal safety.

Competence is my watchword. My two basic responsibilities will always be uppermost in my mind—accomplishment of my mission and the welfare of my Soldiers. I will strive to remain technically and tactically proficient. I am aware of my role as a noncommissioned officer. I will fulfill my responsibilities inherent in that role. All Soldiers are entitled to outstanding leadership; I will provide that leadership. I know my Soldiers and I will always place their needs above my own. I will communicate consistently with my Soldiers and never leave them uninformed. I will be fair and impartial when recommending both rewards and punishment.

Officers of my unit will have maximum time to accomplish their duties; they will not have to accomplish mine. I will earn their respect and confidence as well as that of my Soldiers. I will be loyal to those with whom I serve; seniors, peers, and subordinates alike. I will exercise initiative by taking appropriate action in the absence of orders. I will not compromise my integrity, nor my moral courage. I will not forget, nor will I allow my comrades to forget that we are professionals, noncommissioned officers, leaders!

Appendix II
Various approaches to leadership

The Twelve Principles of Leadership [1]

1. Be technically and tactically proficient. Effective leaders are smart. They not only possess domain knowledge of their field of expertise and their own duties and responsibilities but they know those of their team as well. Your proficiency will earn the respect of your team. Additionally, leaders must understand how best to employ the equipment and weaponry of their team in order to achieve success. This principle demands that leaders take responsibility for staying abreast of current military developments through training, experience, reading and study.

2. Know yourself and seek self-improvement. Professional development is a continuous process. An effective leader evaluates his strengths and weaknesses. Through self-evaluation, a leader is able to recognize his strengths and weaknesses. In this way he can determine his particular capabilities and limitations. As a result, he can take specific actions to further develop his strengths and work on correcting his weaknesses. This process builds self-confidence. An accurate and clear understanding of yourself and a comprehension of Group behavior will help you determine the best way to deal with any given situation. Leaders must be genuine. As Hawthorne so poignantly illustrates, "No man, for any considerable period, can wear one face to himself, and another to the multitude, without finally getting bewildered as to which may be the true." Be genuine and honest.

---

[1] Leadership is the process of influencing people by providing purpose, direction, and motivation while operating to accomplish the mission and improving the organization (ADRP 6-22, 2).

3. Seek responsibility and take responsibility for your actions and decisions. Achieving organizational results means accepting responsibility. Responsibility is demonstrated by decisiveness in times of crisis – not hesitating to make decisions or to act to achieve operational results. Seeking responsibilities also means taking responsibility for your actions. You are responsible for all your unit does or fails to do. Be willing to accept justified and constructive criticism.

4. Know your men and look out for their well-being. Leaders must know and understand those being led. When your men trust you, they will willingly work to accomplish any mission. Successful leaders know their men and how they react to different situations. This knowledge can save lives. Further, knowledge of your men's personalities will enable you, as the leader, to decide how best to employ each man.

5. Keep your men informed. Your subordinates expect to be kept informed, and when possible, have the reasons behind the requirements and decisions explained to them. Your men will perform better and, if knowledgeable of the situation, can carry on without your personal supervision. Providing information inspires initiative.

6. Make sound and timely decisions. Leaders must be able to reason under the most critical conditions. Rapidly estimate a situation and make a sound decision based on that estimation. There's no room for reluctance to make a decision. As MacArthur reminds "never give an order that can't be obeyed." Likewise, subordinates respect the leader who corrects his mistakes immediately.

7. Employ your team in accordance with its capabilities. Successful completion of a task depends upon how well you know your men's capabilities. Good leaders seek out challenging tasks for their subordinates that they are not only

prepared for, but have the ability to successfully complete the mission as well.

8. Ensure each task is understood, supervised, and accomplished. Good leaders communicate their instructions in a clear, concise manner, and allow their men a chance to ask questions. Leaders should also check progress periodically to confirm the assigned task is properly accomplished. This allows your men to know you are concerned about mission accomplishment as well as them.

9. Develop a sense of responsibility in your subordinates. Good leaders show subordinates they're interested in their welfare by giving them the opportunity for professional development. Assigning tasks and delegating authority promotes mutual confidence and respect between the leader and the men. The key element of trust, the fundamental bedrock in the relationship between the leader and the led, is fostered in this way.[2]

10. Set the example. No aspect of leadership is more powerful. Set the standards for your men by personal example. Your subordinates all watch your appearance, attitude, physical fitness and personal example. If your personal standards are high, then you can rightfully demand the same of your men. Your personal example affects your subordinates more than any amount of instruction or form of discipline.

11. Build a team. A good leader is socially astute, and understands that one must train your element with a purpose and emphasize the essential elements of teamwork and realism. Good leaders foster a team spirit that motivates subordinates to work with confidence and competence. As Clausewitz observes, "It is only by means of a great directing spirit that we can expect the full power latent in the troops to

---

[2] Kolenda, *Leadership: The Warrior's Art*, xxii.

be developed."[3] Further, a good leader should always ensure that subordinates know their positions and responsibilities within the team framework.

12. Create a positive climate.[4] Good leaders understand that they set the moral and psychological thermostat in both battle and the workspace. This principle is relative to what is referred to in the military as 'presence.' This principle reminds us that good leaders galvanize the strengths of others. Thus, rather than seeing strengths in others a threat to our leadership, good leaders envision how to implement those strengths and motivate their subordinates to use them. Moreover, control and order are immediately reestablished by presence.[5] A critical component of presence is composure, that is, a leader's command over himself.

---

[3] Clausewitz, *On War*, I. 3.
[4] Historically there have been eleven principles. I (Paul LeFavor) have presumed to add this twelfth principle for the consideration of the military community.
[5] Taylor, *Military Leadership*, 36.

CSM Kevin Dorsh's Notes on leadership[1]

CSM Dorsh – Who I am and what I believe: I am an introvert. I enjoy talking one on one versus in a big Group. Relationships on a personal level matter to me.
Family matters to me (I have learned from my mistakes as a 1SG/CO SGM – heed that warning, don't make the same mistakes I did!). Our fallen and their families are important to me.

I am in my head a lot which makes me appear as unapproachable at times (you say resting bitch face, I say stoic). Don't judge a book by its cover – it's usually because I am thinking about something. I have a horrible memory (working on that through WRC/THOR). If you tell me something and I don't write it down – it's gone. Send me an email to follow up. I will ask you to do this from time to time – this is why.

PT is important to me. Please feel free to say hi to me in the morning at THOR. Let's not talk shop during PT. Let's talk about our families, personal hobbies, etc. – work can wait for the duty day. PT is our time to work on ourselves.

Counseling/Personal Development is important to me. A positive mindset is important to me – it's infectious to an organization.

Competition is important to me. I am open to and expect spirited debate from our CSM/SGMs. I do not have all the

---

[1] CSM Kevin Dorsh was the 10th Special Forces Group's CSM when this book was written (2022). Although some of his Notes are pointed at Company, BN, and GRP level leaders, most are 100% usable at the ODA level.

answers. I will make mistakes, you will too! I have been in 10th Group for 20 years straight; I have my biases – I understand that and am working on that. I have had a long stare at common problems within our organization – just as some of you have. With your help, we can make a difference. I am deeply passionate about seeing 10th Group being the Best Organization it can be for our SMs.

1) Communicate, Communicate, and Communicate!! Don't take it for granted and assume that since you put it out in a meeting that it reached the lowest levels. Constantly check. Figure out if there is a break in the chain...maybe it is you? A lot of mental calories are burned over miscommunication. I am at fault for this too. IF we are deliberate and disciplined in our commo, we can fix this issue in Group.

2) Strive to find harmony between work and family. It is not a balance, its harmony. What works for me and my family, may not work for you and yours. Find what works for you. Create a contract with your family (talk about it with them) and then fight to stay true to it.

3) Be the first adult in the room. Make tough decisions. Ask the tough questions. Don't take the easy way out and just rubber stamp a recommendation IOT force your next higher level of leadership to make the tough decisions for you. Think 2nd and 3rd order effects. Does your decision support the Group's values?

4) You are now past the tactical level of leadership and are moving to the **Organizational level.** Understand that. Work well with you peers to you left and right at echelon. Create a sense of community within your organization. Is it for the greater good – the Organization? Or is it for YOUR greater good? Us vs. Them mentality – be careful, that can be polarizing. Never talk bad about your higher HQ.

5) Analyze your training calendar. What are you putting your resources against (time/$)? Are we devoting time/$ to personal development of our up-and-coming leaders (you)? We say we are doing it, but are we? Your calendar will tell me if you are. We have train/man/equip responsibilities while here at Carson/Panzer. Leader Development of our personnel is part of that. We tend to focus on operations – because that's easy – that's what we know. Training leadership is hard – find a creative way to do it.

6) Fight complacency. It's human nature to take the path of least resistance. Strive to get better every day. Put yourself in uncomfortable situations. Force yourself to learn something new every day.

7) Do the internal work. Understand who you are as a person and your core values. This will highlight potential insecurities, areas where you are not comfortable working with. The first task is to identity an area that needs work, then work on it – always strive to get better. Use the assets we have afforded us within the Group and Army. In times of crisis, the "real" you will come out – know what that is.

8) Write a core statement. Put it up on your board in your office where you and others will see it every day (Accountability). Here is mine: ***Relentlessly strive to be present and genuine in daily life by living passionately for my family, my work, and myself.***

9) Incorporate a strict and disciplined work week calendar: Monday – prep/admin, Tuesday through Thursday – train day and night, Friday – recovery/admin. It's simple to understand and implement and proven to work. Take a look at when you are holding Company/Battalion training meetings/Command and Staff – is it Tuesday – Thursday? What message does that send to the force and to our leaders at echelon?

10) Training – Always work to incorporate a psychological and physical component to training. **Work to inoculate stress into your formations.**

11) Competition Breeds Success – Man in the Arena – Tomorrow is not a right; earn it! Those are my mantras – What are yours? **Does your video match your audio?** You are moving into the realm where everybody is watching what you do and listening to what you say. Be conscious of that.

12) Do you know the **Group values**? Do you live by them? Trust, Excellence, Commitment, Character, Accountability. How do we as the Senior Leaders of this organization exhibit these values?

13) The relationship between the Officer and the NCO is critical. Development of our Officer counterpart (Reverse Mentorship) is arguably one of the most important duties as Team Sergeant/Company Sergeant Major. Don't isolate, but work to insulate our officers. Show them how to use an NCO (trusted and reliable) at all levels.

14) Treat your SFRG as another tactical maneuver element. You provide guidance, training, and resources to your maneuver elements. Do the same for you SFRG. If you train them, resource them = they can be counted on in times of crisis.

15) Physical Training: Don't workout, train. Physical Training is a key component of your element's ability to successfully complete their assigned mission. Have a plan, work towards a specific end state. Use THOR whenever possible.

16) Create a culture where people WANT to be tested. Conduct Bee Stings, Company and Battalion evaluation events.

Embrace the Group Commander's EDRE policy. Prepare your teams - readiness is not just a work or function, it is a mindset.

17) Conduct PT as a Team/Element as often as possible. Through shared adversity, a team will bond and create a sense of resiliency amongst themselves. Never stop trying to build the Team no matter what level you are at.

18) Fight. Get your teams/elements in the Dojo. We no longer fight as a Group. There is dust on our mats in the Dojo. What does that say about our culture?

19) Devote time and energy to the development of your Soldiers. Don't conduct **Random Acts of Development** – ensure that your leadership program builds upon itself and has a defined end state.

20) Don't try to find the **"right time"** to conduct LPDs IOT make sure everyone is present. That will never happen and it will continue to be pushed to the right. Schedule it, put it on the training calendar at echelon, and execute – they will come. Think about scheduling on Monday or Friday.

21) Don't manage from your Office/Desk. Be seen. Walk around your formations. Be approachable. I will be in your spaces. I will be walking into Team Rooms. Don't worry about it. It's my way of building relationships with the guys. Do the same.

22) Strive to create an environment so that we don't have the **meeting after the meeting**. We can have a tough, honest, candid, and transparent conversations amongst ourselves as CSMs and SGMs. It's not personal, it's professional. We owe transparency to our formations and amongst ourselves.

23) Positive Intentions Always (PIA). Believe that your Soldiers in Group always start out with the best intentions. I believe the same about you. Try to approach all situations with a positive attitude.

24) NCOERs/Awards are important. One of the few things that will follow our SMs for the rest of their lives. Company SGMs/1SGs – **EVERY NCOER/AWARDS** needs to be QA/QC'd through you. CSMs – Actively manage your system. Do a 10% check quarterly. Own and manage your E8/Team Sergeant population. All E9 Evaluations will run through the Group CSM. Trust but verify.

25) Reinforce the Group's Enlisted Manning Cycle – provide predictability to our SMs. Soldiers can expect to PCS to SWC/OUTSOC or 1-10 at the 48–60-month mark. Coming from 1-10, they can expect to PCS at the 36-month mark to SWC or 10th Main. 1-10 Soldiers can extend in Germany to 60 months (5 years), no questions asked. **Manning 1-10 is important to me.**

COL. Glover S. Johns'[1] Philosophy of Soldiering

COL Glover S. Johns was an infantry officer who stormed the beaches of Normandy with the 29th Division and fought all across Europe. He also fought in Korea as a XO, and Regimental Commander. He authored a book, *The Clay Pigeons of St. Lo,* chronicling his Task Force's capture of the critical French town of St. Lo. He was highly regarded by COL (ret) Hackworth in his book *About Face.*

1) Strive to do the small things well.
2) Be a doer and a self-starter—aggressiveness and initiative are two most admired qualities in a leader—but you must also put your feet up and *think*.
3) Strive for self-improvement through constant self-evaluation.
4) Never be satisfied. Ask of any project, *how can it be better*?
5) Don't over-inspect or over-supervise. Allow your leaders to make mistakes in training so they can profit from the errors and not make them in combat.
6) Keep troops informed; telling them "What, how, and why" builds their confidence.
7) The harder the training, the more troops will brag.
8) Enthusiasm, fairness, and moral and physical courage—four of the most important aspects of leadership
9) Showmanship—a vital techniques of leadership.
10) The ability to speak and write well—two essential tools of leadership.
11) There is a salient difference between profanity and obscenity; while a leader employs profanity (Tempered with discretion), he never uses obscenities.
12) Have consideration for others.

---

[1] COL Glover Johns was a mentor to COL (ret) David Hackworth, these principles appear in his book, *About Face*, page 402.

13) Yelling detracts from your dignity; take men aside to counsel them.
14) Understand and use judgment; know when to stop fighting for something you believe is right. Discuss and argue your point of view until a decision is made, and then support the decision wholeheartedly.
15) Stay ahead of your boss.

Appendix III

Tactical Leadership, Chapter 1: Philosophy of Leadership, BLEMISH

Below I straight up stole (ok, I had permission) Paul Lefavor's full explanation of his acronym BLEMISH.

B – Blame Shift: Toxic leaders blame others to cover up mistakes. As Norman Dixon points out, blame shifting is a psychological defense mechanism whereby leaders or organizations by denial, rationalization, or by making scapegoats (or a mixture of the three) use to deflect blame. Dixon observes, "However it is achieved, the net result is that no real admission of the failure or incompetence is ever made by those who are really responsible."[1]

L – Lackey: This is the 'people-pleasing' leader. Sure, your people should love to work for you, but leadership is not democracy. This can still be toxic because people-pleasing leaders tend to avoid confrontation and overlook problems for the sake of popularity. By avoiding too great familiarity with the men, they will not only fain their love and confidence, but be treated with a proper respect, whereas by a contrary conduct they forfeit all regard, and their authority becomes despised.

E – Egotistical: Egotistical people are conceited, self-absorbed and self-centered. Toxic leaders take credit for other's

[1] Norman F. Dixon, *On the Psychology of Military Incompetence* (Philadelphia: Basic Books, 1976), 32.

success. These are the backstabbers. Mature leaders give credit where credit is due. Leaders who fall into this pitfall envision their current position as a means for merely advancing their own career. However, leadership doesn't mean you use others to advance your own career. Egomaniacs are toxic leaders. They often disrespect their people and treat them poorly. Exceptional units are the ones who are led by officers and NCOs who understand the professional relationship between their people and themselves. They treat their people with respect and honor.

M - Micromanage: Toxic leaders sap the energy out of self-starters. Micromanagement occurs when leaders who don't trust their people, closely control work, planning, etc. This eventually leads to a breakdown of trust between leaders and subordinates. Rather than having the confidence to allow subordinates to make decisions on their own, micromanagement says, "I don't trust you." All leaders would do well to remember Patton's leadership axiom that denounces micromanagement: "Tell people what to do, not how to do it."

I – Inability to keep your word: Toxic leaders don't keep their word. Effective leaders don't make promises they can't keep. What causes a leader to make things sound better than they really are is a desire for popularity. Leadership is not a popularity contest. Straight shooters are respected more for their candor. Honesty, as they say, is always the best policy.

S – Self-Control (lack of): A man who would wisely govern others must be able to govern himself. Charles Martin observes, a leader "who cannot control his emotions of anger, excitement, etc., or who is swayed by his impulses of vanity, egotism, ambition, or personal prejudices, cannot obtain the best results from others, nor give his own best service to the cause."[2] Likewise Clausewitz observes, "Strength of character does not consist solely in having powerful feelings, but in maintaining one's balance in spite of them." In short, a leader without self-control loses the confidence of those he leads.

H – Hypocrisy: For a leader to expect any measure of respect from his followers he must avoid being hypocritical. Woe to the new leader who thinks he knows everything! Greater still is the leader who in his naivete puts on an arrogant false face, pretending to be someone he's not. Hawthorne put it best: "No man, for any considerable period, can wear on face to himself, and another to the multitude, without finally getting bewildered as to which may be true."[3]

---

[2] Charles F. Martin, *Winning and Wearing Shoulder Straps* (Whitefish, MT: Kessinger Publishing 2010), 57.
[3] Nathaniel Hawthorne, *The Scarlet Letter* (New York: Heritage Press, 1935), 231.

Appendix IV
My Initial Counseling Form

Key points of discussion:
Outlined below are my expectations of you as the Team Sergeant of Operational Detachment Alpha 0332:

1. LEADERSHIP: You are the heart and soul of ODA 0332. This team will take on your personality, and the NCOs will emulate your work ethic and leadership style. Your responsibilities to this team are insurmountable, but with the properly directed effort you will be extremely successful. You are the standard enforcer, which may make you unpopular at times, accept that fact now. You must be a strict disciplinarian. The "little things" (punctuality, laziness, minor discipline infractions, attitudes...etc.) are the easiest to let slip by, but will always come back to haunt the Team. Delegate everything.

2. DEDICATION: Your time on a Special Forces Detachment is a very small fraction of your life, but one of the most defining. As the Detachment's Team Sergeant, the Team's success or failure depends on your actions, planning, and dedication. Make this short period of time one of the most productive of your life, by truly dedicating your full efforts towards this team. The men on you team deserve it, and you deserve the success.

3. PHYSICAL FITNESS: ODA 0332 must be EXTREMELY physically fit in order to successfully complete our mission as a Mountain Team. The Team's leadership MUST set the example for all ODA NCO's when it comes to physical fitness. This does not necessarily mean that the leadership is required to be the fastest, or strongest (considering our environment, this would be rare). However, the leadership must consistently demonstrate the most effort, and determination to maintain their fitness. In a nutshell, work harder than everyone else.

4. COMMUNICATION: You must always disseminate information to the team in a timely manner. This requires some kind of system to ensure that everyone remains informed. We are a leadership team, which also requires constant cross-talk and positive communication. We will both be involved in major decisions, although I will always trust your judgment, I hope you would never hesitate to unilaterally make a major decision in my absence. I will keep you informed of all of my thoughts and plans pertaining to the team, and I expect you to do the same.

5. TRAINING: You are the primary trainer of this Detachment. The training schedule and focus is your responsibility. I will give you my input pertaining to the long-term training schedule, as I will always keep my focus on the broader, long-term issues. I count on you to keep the team focused on all day to day, week to week training. With our specific mountain specialty, our ODA should never lack anything to do. We could always spend our empty days maintaining proficiency in LVL III and LVL II tasks.

6. MAINTAINING: You are responsible for the maintenance of all team equipment.

7. OPERATING: Although we are not slated for a combat deployment in the foreseeable future, we are a Special Forces Team, and therefore always on a "beeper". As the premier mountain detachment in the regiment, we must be prepared at all times to conduct combat in the most austere mountainous environment.

Annex I

NOTE: Approved for use by SWCS CG/Staff on 24 August, 2023.

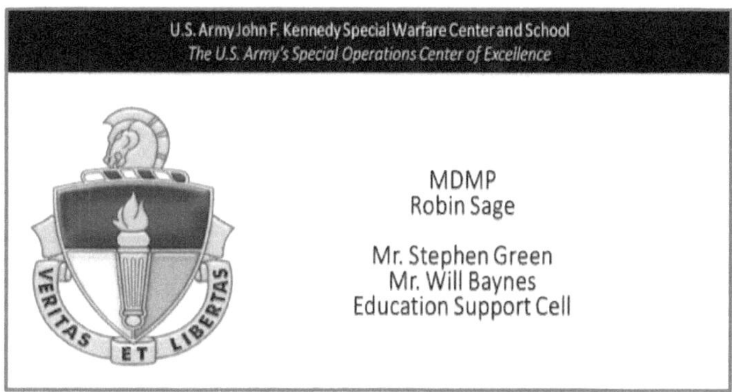

PURPOSE- The purpose of this paper is to assist SWCS cadre in understanding MDMP doctrine, identify best practices and apply Special Operations missions to MDMP where doctrine does not specifically guide SOF mission planning. This paper will not cover all the steps to MDMP in depth, only areas identified as friction points and topics based on observations. Comments made in *italics* are taken straight from doctrine. This paper is specifically written for D Co 1/1 SWTG cadre. When understanding or teaching MDMP, instructors should first understand Army doctrine, specifically the basic seven steps to MDMP (see references below, specifically FM 6-0)

References (**current during the creation of this paper**):

GTA 31-01-003 SF Detachment Mission Planning Guide (Jul 2012) (This GTA may not follow current doctrine).

ADP 1-02 Operations Terms and Graphics (these pubs update almost yearly, previous editions conflict with current edition) (Aug 2018)
ATP 2-01.3 Intelligence Preparation of the Battlefield (Mar 2019)
ATP 3-21.10 The Infantry Company (Jul 2016)
FM 3-90-1 The Offense and Defense (Mar 2013)
FM 3-18 Special Forces Operations (May 2014)
TC 3-21-76 Ranger Handbook (Apr 2017)
ADRP 3-0 Operations (Nov 2016)
ADRP 5-0 The Operations Process (good planning pub) (May 2012)
ADRP 6-0 Mission Command (May 2012)
FM 6-0 Commander and Staff Organization and Operations (CH 9 is MDMP) (May 2014 w/updates)

Functioning similar to a battalion staff for planning, an SFODA uses MDMP. MDMP has seven steps (see figure below):

Key Inputs	Steps	Key Outputs
• Higher headquarters' plan or order or a new mission anticipated by the commander.	Step 1: **Receipt of Mission**	• Commander's initial guidance. • Initial allocation of time.
	Warning Order	
• Commander's initial guidance. • Higher headquarters' plan or order. • Higher headquarters' knowledge and intelligence products. • Knowledge products from other organizations. • Army design methodology products.	Step 2: **Mission Analysis**	• Problem statement. • Mission statement. • Initial commander's intent. • Initial planning guidance. • Initial CCIRs and EEFIs. • Updated IPB and running estimates. • Assumptions. • Evaluation criteria for COAs.
	Warning Order	
• Mission statement. • Initial commander's intent, planning guidance, CCIRs, and EEFIs. • Updated IPB and running estimates. • Assumptions. • Evaluation criteria for COAs.	Step 3: **Course of Action (COA) Development**	• COA statements and sketches. - Tentative task organization. - Broad concept of operations. • Revised planning guidance. • Updated assumptions.
• Updated running estimates. • Revised planning guidance. • COA statements and sketches. • Updated assumptions.	Step 4: **COA Analysis (War Game)**	• Refined COAs. • Potential decision points. • War-game results. • Initial assessment measures. • Updated assumptions.
• Updated running estimates. • Refined COAs. • Evaluation criteria. • War-game results. • Updated assumptions.	Step 5: **COA Comparison**	• Evaluated COAs. • Recommended COAs. • Updated running estimates. • Updated assumptions.
• Updated running estimates. • Evaluated COAs. • Recommended COAs. • Updated assumptions.	Step 6: **COA Approval**	• Commander approved COA and any modifications. • Refined commander's intent, CCIRs, and EEFIs. • Updated assumptions.
	Warning Order	
• Commander approved COA and any modifications. • Refined commander's intent, CCIRs, and EEFIs. • Updated assumptions.	Step 7: **Orders Production, Dissemination, and Transition**	• Approved operation plan or order. • Subordinates understand the plan or order.

Legend:

CCIR commander's critical information requirement
COA course of action
EEFI essential element of friendly information
IPB intelligence preparation of the battlefield

Where to Go in Doctrine for Planning Questions?

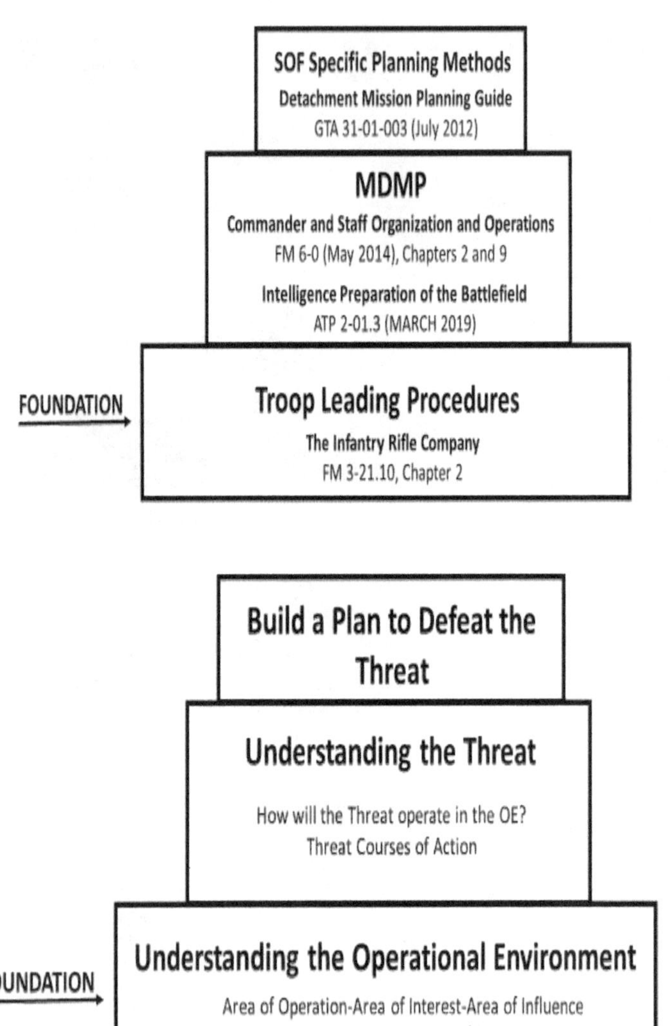

MDMP vs TLPs

There are many similarities between TLPs and MDMP. The MDMP methodology integrates the activities of the Commander, staff, subordinate headquarters, and other partners to:

- Understand the situation and mission.
- Develop and compare COAs.
- Decide on a COA that best accomplishes the mission.
- Produce an operation plan (OPLAN) or operation order (OPORD) for execution.

TLPs on the other hand extend the MDMP to the small-unit level. The MDMP and TLP are similar but not identical. Troop leading procedures are a dynamic process used by small-unit leaders to analyze a mission, develop a plan, and prepare for an operation (ADP 5-0). These procedures enable leaders to maximize available planning time while developing effective plans and preparing their units for an operation. (ATP 2-01.3 IPB).

The standard Army planning process has five interrelated subprocesses: mission analysis, course of action (COA) development, COA analysis, COA comparison, and COA selection. (FM 3-21-10)

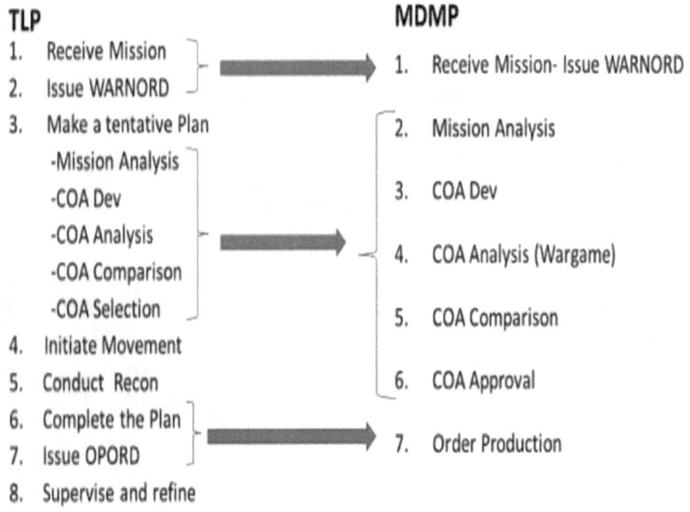
TLP vs. MDMP
TLP
1. Receive Mission
2. Issue WARNORD
3. Make a tentative Plan
-Mission Analysis
-COA Dev
-COA Analysis
-COA Comparison
-COA Selection
4. Initiate Movement
5. Conduct Recon
6. Complete the Plan
7. Issue OPORD
8. Supervise and refine
MDMP
1. Receive Mission- Issue WARNORD
2. Mission Analysis
3. COA Dev
4. COA Analysis (Wargame)
5. COA Comparison
6. COA Approval
7. Order Production

IPB

Despite any differences, the one thing that remains critical to any plan is Intelligence Preparation of the Environment (IPB). ATP 2-01.3 states "IPB results in the creation of intelligence products that are used during the MDMP to aid in developing friendly courses of action (COAs) and decision points for the Commander. Additionally, the conclusions reached and the products created during IPB are critical to planning information collection/intelligence collection and targeting operations (para 1-2). If planners break that statement down, they can deduct that as a Commander it is difficult to make a decision without conducting proper IPB, it is the foundation to the plan. Additionally, it is an ongoing and never-ending process.

It is also important to note that during the planning and operations process planners should no longer refer to the adversary as the "ENEMY". Instead doctrine now refers to them as the threat/adversary. Using the term "enemy" leaves planners with a vision of a uniformed and organized military. This is especially important in SOF operations since SOF operators often work in a very ambiguous and undefined environment. Using the term threat/adversary implies including any element that may pose a threat to your operations. A threat is any combination of actors, entities, or forces that have the capability and intent to harm United States forces, United States national interests, or the homeland. Threats may include individuals, Groups of individuals (organized or not organized), paramilitary or military forces, nation-states, or national alliances. Commanders and staffs must understand how current and potential threats organize, equip, train, employ, and control their forces (ADRP 3-0).

IPB is broken down into 4 steps:

1. Define the Operational Environment
2. Describe the environmental effects on operations
3. Evaluate the threat
4. Determine Threat COAs

*** NOTE- When using these tools there is more to the analysis than just the facts. Analysis should include the "so what" and even the "now what". Planners identify a fact, analyze that fact and the effects on friendly and threat, and then take action. In other words, what are the opportunities, not just for the threat but also for us?

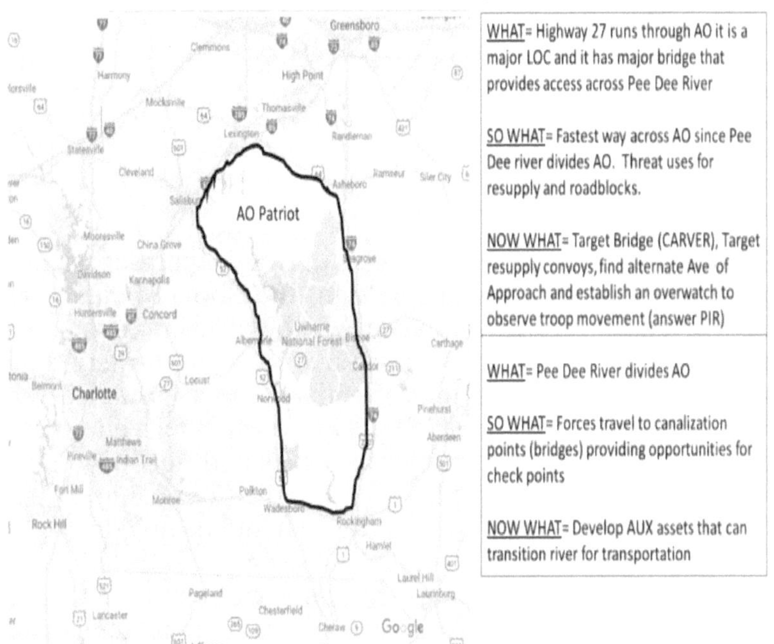

Step 1- Define the Operational Environment.
An operational environment is a composite of the conditions, circumstances, and influences that affect the employment of capabilities and bear on the decisions of the Commander (JP 3-0). An OE for any specific operation comprises more than the interacting variables that exist within a specific physical area. It also involves interconnected influences from the global or regional perspective (such as politics, economics) that affect OE conditions and operations. Thus, each Commander's OE is part of a higher Commander's OE.

Understanding friendly and threat forces is not enough; other factors, such as culture, languages, tribal affiliations, and operational and mission variables, can be equally important. (ATP 2-01.3)

Define the operational environment is broken down into 3 sub steps:

1. Area of Operations
2. Area of interest
3. Area of influence

Area of Operations (AO)- This is the unit's specific assigned boundaries that they will operate in, NOT broad boundaries (i.e. the entire JSOA). Assigned AOs by doctrine "should be large enough to accomplish their missions and protect their forces (JP 3-0)". By assigning specific AOs this allows the unit to analyze the AO they will operate in, which in turn allows the Commander to make informed decisions and not waste time focusing on areas that are irrelevant to the Commander. AOs should also be flexible. Commanders should be able to adjust their AO based on their analysis. For example, if the assigned AO splits a city or area that might represent a center of gravity (COG) and the Commander feels he cannot operate in or influence just half, then he might consider requesting AOs be adjusted so the entire COG is in his AO.

**A best practice for briefing the AO is to start broad and work in. To do this remember OBTF:

- <u>Orient= Broad</u>. This is just getting a reference point on a map, this could include major cities, states, or terrain features (like the Appalachian Mountain range)
- <u>Box= General</u>. There are a few techniques from using major avenues of approach to grid lines or borders. The preferred method would be to use something you can physically see on the ground
- <u>Trace=specific</u>. This is the specific borders that constitute your AO

- <u>Familiarize</u>= Major features inside your AO that are easily recognizable on the ground, this should provide an initial step to terrain analysis and a starting point for the Commander's guidance

EXAMPLE Brief:
Orient= "To the north we have Greensboro NC, to the west we have Charlotte, NC, to the South we have Florence SC, to the east we have Fayetteville, NC"
Box= "To the north we have interstate 40 running east west, to the west we have interstate 77 running north south, to the south we have highway 74 running east west, and to the east we have Highway 1 running generally north south"

Trace= "Our AO runs from Lexington generally southwest along interstate 85 to highway 52 then south along highway 52 to Wadesboro then east on highway 74 to interstate 74 then north to highway 64 at Asheboro then west to Lexington"
Familiarize= "In our AO we have the Uwharrie National Forest in the center, the Pee Dee River running north to south on the west side of the AO and includes Lakes Tillery, Badin and High Rock. Dividing our AO we have highway 27 that runs from the city of Biscoe on the east side of the AO to the City of Albemarle on the west side of the AO."

Area of Interest (AOI)- This is the area generally outside a Commanders AOI that influences inside their AOI. It is commonly referred to as outside in. AOI is that area of concern to the Commander, including the area of influence, areas adjacent thereto and extending into enemy territory. This area also includes areas occupied by enemy forces who could jeopardize the accomplishment of the mission (ATP 2-01.3). When planners analyze our AOI there are many factors that they must consider. Commonly when conducting IPB for TLPs, the term CAR (**C**AS/CCA, **A**rtillery/Mortars, **R**eserves/Reinforcements) is used to analyze those factors influencing our AI. However, when conducting MDMP planners must consider other factors in their AOI. These factors could be political, ideological, Business, money, other threat Groups, crime, other state or non-state sponsors, etc. Commander can also divide the AOI into components (air/ground). Ideally keep this to a manageable number, around 3-5. This allows the Commander to focus on what is most important and it is easy for Soldiers and staff to remember. When briefed, AOI analysis should focus on how those factors will impact operations in AO.

The 3-5 rule should be used throughout the planning process for all planning tasks.

EXAMPLE- The Iranian influence in Iraq especially in predominately Shiite controlled areas. This influence could be projected to areas as large as whole cities (i.e. Najaf) or as small as neighborhoods or ethnic enclaves (Sadr City in Baghdad).

Another common mistake is to analyze factors already in the AO, this will happen later on in IPB. Concentrate only on factors outside the AO.

Area of Interest – "Outside In"

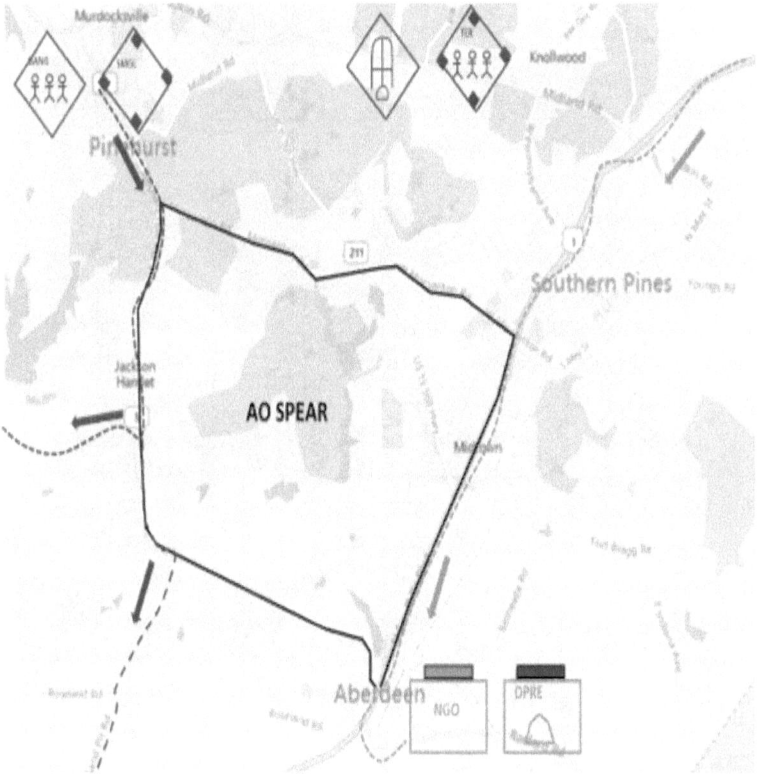

Area of Influence (AI)- An area of influence is a geographical area where a Commander is directly capable of influencing operations by maneuver or fire support systems normally under the Commander's command or control (JP 3-0). This is commonly referred to as the inside out. However, ATP 2-01.3 states AI is "an area that includes terrain inside and outside the AO". Like AOI, there are more considerations taken other than the CAR factors. Again, keep this to no more than 3-5.

For example, if the Commander influences a radio tower that is located in their AO but the tower has the ability to transmit beyond the assigned AO into adjacent AOs, then that tower's transmit area would be part of the Commander's AI.

Area of Influence – "Inside Out"

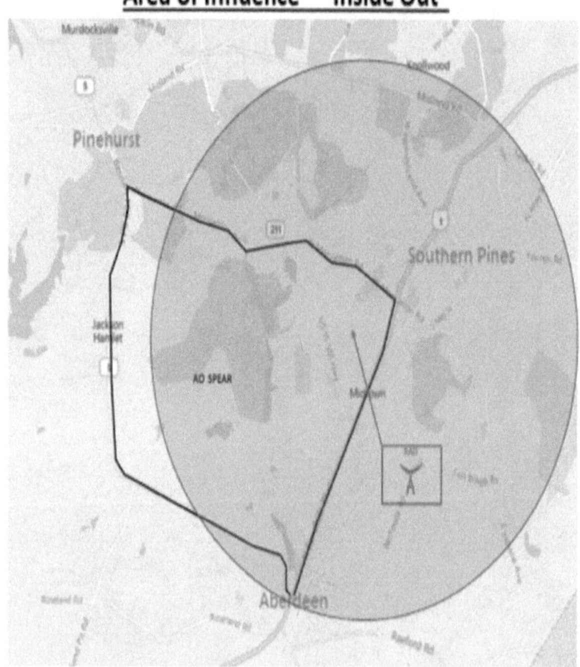

**NOTE- Understanding your Operational Environment is also an ARSOF imperative- "SF cannot shape the operational environment without first gaining a clear understanding of the joint operations area, including civilian influence and enemy and friendly capabilities. SF applies the political, military, economic, social, information, infrastructure, physical environment, time operational variables to analyze the operational area. SF must identify the friendly and hostile decisionmakers, their objectives and strategies, and the ways they interact. The conditions of conflict can change and SF must anticipate these changes in the operational environment and exploit fleeting opportunities." FM 3-18. However, if you correlate the ARSOF attribute to IPB you will see that the attribute is essentially inclusive to all the steps of IPB.

GTA 31-01-003-page 2-11 step 1 and the IPB manual (ATP2-01.3) states identifying or describing significant characteristics of the AO are part of step 1 of IPB. During the FAMILIARIZE portion of the AO brief, the Commander has identified those, from there he should give GUIDANCE for further analysis.

Step 2- Describe the battlefield effects
This is where the analysis of previous identified features is accomplished. When describing the battlefield effects, we have several tools that assist us. These tools include:

1. OAKOC
2. WEATHER
3. PMESII-ASCOPE

There should be some sort of graphic to depict the terrain analysis- For MDMP this should be a Modified Combined Ops Overlay (MCOO) and for SUT (TLPs) there should be a Graphic Depiction of Terrain (GDOT). The preferred method for the terrain overlays is an analog product (map and plastic overlay). The reason for this is time. It takes a lot of time to develop map overlays using power point.

After the analysis is complete the entire staff should view the product. In the case of an ODA the entire team should look at the product by MOS and determine how the analysis pertains to their MOS. For example, the 18E would look at the analysis and determine what are the LOS restrictions, likely relay sights and likely DF sights.

OAKOC:

- Observation and fields of fire- observation is the ability to view either visually or through surveillance devices. Weather can affect observation. Fields of fire is an area that a weapon system can affectively cover with fire. If terrain offers good cover then it generally limits fields of fire. This should be analyzed based on the weapons systems capabilities of both friendly and threat organic weapons
- Avenues of Approach (AA)- The most important thing about AA is that in order to be an AA, it has to lead to something (objective, key terrain, etc). A network of roads, or trails that people or forces travel along is defined as a mobility corridor. Mobility corridors turn into AA. AA can be ground or Air. Regardless of being a mobility corridor or an AA, there needs to be analysis on the size and type and how it relates to capability. For example, HWY XX is a 4-lane divided improved highway and can support a Company of mechanized traveling in file formation at 35 miles per hour (this is usually determined based on threat doctrine).
- Key Terrain- Key terrain is usually a point of friction to different planners. However, doctrine defines key terrain as "any locality or area the seizure, retention, or control of which affords a marked advantage to either combatant". Key terrain is often used to determine battle positions or could transition into objectives. Key Terrain can change by phase. In SOF operations it is important to point out that it may be something other than a physical feature- In counterinsurgency operations, key terrain may include portions of the population, such as political, tribal, and religious

Groups or leaders; a localized population; infrastructure; or governmental organizations (2-01.3 para 4-32).

- Obstacles- What are the major obstacles. Obstacles are features that impact operations, it can be natural or man-made. In the example used previously, the Pee Dee River would be an obstacle since it divides the AO and canalizes travel.
- Cover and concealment- Cover protects you from bombs, bullets and fragments. Concealment hides you from observation.

Weather:
When analyzing weather considerations, planners should analyze immediate (10-30 day forecasts), seasonal (Summer is the hottest months and produces droughts), and long term historical weather events (Hurricane season). You can include long term impacts from major weather events (Hurricane Katrina)

PMESII/ASCOPE (AKA Human Terrain Analysis):
The biggest mistake made with Civil factors is it usually stops at the "so what". Analysis should be done not only for the threat but also for friendly/HN/partner forces. Another common mistake made with Civil considerations is the lack of focus that often produces large amounts of information. Planners should instead focus civil considerations based on needs, such as support networks or by MOS/staff sections.

For example, the 18D concentrates his focus on the medical network development. The medical network consists of the point of injury, to the casualty collection point or ambulance exchange point, to the "g" hospital, and finally to a convalesce facility. The 18D can focus his PMESII collection on those requirements.

Commanders should give guidance prior to starting analysis. The S2/18F should manage this and NOT try to produce the product themselves. ODA Planners should also do analysis based on personnel qualifications (if feasible), if you have a planner with an economics degree or background, they would be better suited to understand the economics section (MOSs don't necessarily correlate to section expertise).

Civil considerations, if done effectively, can be used for many factors such as RFI/RFS development, PIR/IR development, and targeting. PMESII/ASCOPE is the best way to determine kinetic and non-kinetic targeting and develop High Value Target List (HVTL) and High Payoff Target List (HPTL) which can include structural (which in turn can be further analyzed using CARVER) and human targets. An effective targeting method when analyzing civil considerations during planning is the D3A method. For tactical execution the F3EAD model works best. The key to D3A is the measure of effectiveness (MOE) during the assessment. There must be some method to assess how the targeting worked and how effective it was. A common mistake is to measure the effectiveness of targeting through a measure of performance (MOP) instead of a MOE.

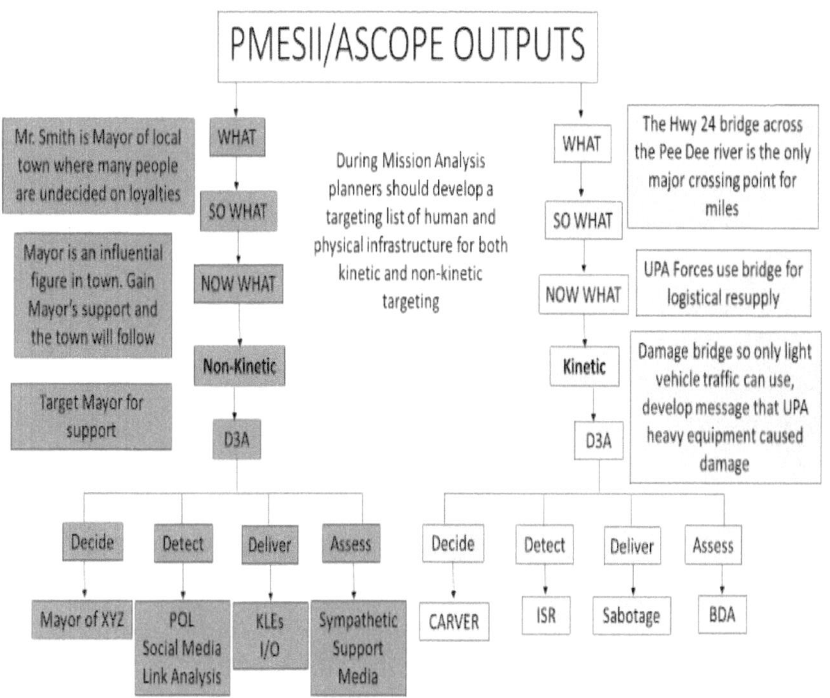

An example of IO targeting using the D3A method:

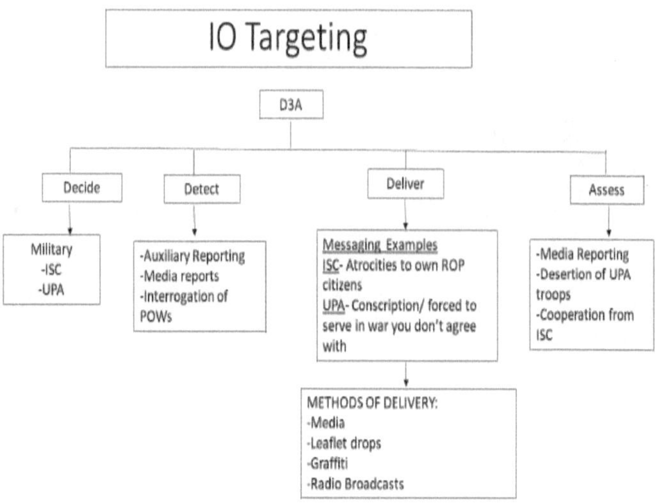

Step 3: Evaluate the threat

Commanders should evaluate their partner force as well. ADP 1-02 Operations Terms and Graphics has symbols for about every threat and friendly unit or organization, action or incident you need. It even has a breakdown of how irregular or hybrid threats may be pictured. This step should consist of:

- Situation/Road to war- who are they and where did they come from (2 levels up, if there is a BN HQ in your AO where is the BDE, DIV HQ)
- Disposition-where are they on the map.
- Composition- Threat identification and how they are organized. ADRP 2-01.3 para 5-24. Composition describes how an entity is organized and equipped—essentially the number and types of personnel, weapons, and equipment. The staff uses line-and-block chart products to visually see the enemy's composition.
- Strength- this is described in term of weapons, personnel and equipment. It is a good practice to use actual numbers (80% strength of 100 troops=80

troops), and what is the so what and possibly the now what (action) of this.

- Capabilities- By War fighting function to include the threats forces human/civil engagement
- Doctrine- how does the threat doctrinally fight or employ forces.

**Note After evaluating the threat there should be a threat overlay. When doing a threat overlay there should be some analysis on the threat TWO levels down. For example, if there is a Battalion size element within the assigned AO, then planners should do some analysis on where the Companies and Platoon might be located based off the threat evaluation. Planners should also be able to make or add to an HVT list from the analysis

<u>**Step 4**- Determine Threat COAs</u> ** NOTE- there was a change to doctrine on threat COAs
Based on the results of the mission variables analysis conducted earlier in the IPB process, the staff now identifies

the threat's likely immediate and subsequent objectives and desired end state. These elements are included in the threat COA statement developed for each COA.

To plan for all possible contingencies, the Commander understands all COAs a threat Commander can use to accomplish objectives. The staff assists in this understanding by determining all valid threat COAs and prioritizing them from most likely to least likely. The staff also determines which threat COA is the most dangerous to friendly forces. To be valid, threat COAs should be feasible, acceptable, suitable, distinguishable, and complete—the same criteria used to validate friendly COAs.

Generally, threat forces are more likely to use a COA that offers the greatest advantage while minimizing risk. However, based on the situation and its objectives, the threat may choose to accept risk to achieve a desired end state. It is impossible to predict what COA the threat will choose. Therefore, the staff develops and prioritizes as many valid threat COAs as time allows but, at a minimum, develops the most likely and most dangerous COAs.

1. Most probable course of action- What the threat is doing now and will most likely do in the future
2. Most Dangerous course of action- What is most dangerous to friendly forces. The most important aspect to the MDCOA planners must consider is how much risk the threat Commander is willing to take, is it within their capability and doctrine, and is it most dangerous to you.

Upon identifying all valid threat COAs, the staff compares each COA to the others and prioritizes them by number. For example, if four COAs have been developed, COA 1 is the threat's most likely COA, and COA 4 is the least likely. Additionally, the staff determines which COA is the most dangerous; however, the designation of the most dangerous COA largely depends on how much each threat COA threatens the selected friendly COA. The most likely COA may also be the most dangerous. Additionally, a COA needs to answer six basic questions:

- **Who** (the organizational structure of the threat organization, including external organizations providing support)?
- **What** (type of tactical mission task such as defeat, destroy, seize)?
- **When** (the earliest time the action can begin)?
- **Where** (the battlefield geometry that frames the COA [boundaries, objectives, routes, other])?
- **How** (the threat attacks, defends)?
- **Why** (the threat's objectives)?

After identifying the full set of potential threat COAs, the staff develops the tools necessary to determine which COA the threat may implement. Because the threat has not acted yet, this determination cannot be made during IPB. However, the staff can develop the information requirements and indicators necessary to support the construction of the information collection plan that can provide the information necessary to confirm or deny threat COAs and locate threat targets.

Information requirements are, in intelligence usage, those items of information regarding the adversary and other relevant aspects of the operational environment that need to be collected and processed in order to meet the intelligence requirements of a Commander (JP 2-0). An indicator is, in intelligence usage, an item of information which reflects the intention or capability of an adversary to adopt or reject a course of action (JP 2-0). Identifying and monitoring indicators are fundamental tasks of intelligence analysis, as they are the principal means of avoiding surprise. Indicators are often described as forward looking of predictive indicators. (ATP 2-01.3 Ch. 6)

NO WHERE IN the IPB manual does it state "The threat COAs CAN NOT be a result of friendly action." This is one area that it is only found in the GTA.

To sum this up: Planners should develop as many threat COAs as time permits. Prioritize them, determine which is the most likely and the most dangerous (they could be the same), develop a Threat COA statement and sketch (for each), then develop IR/Indicators for each COA.

PLANNING

KEY COMPONENTS OF A PLAN

ADRP 5-0 para 2-69. The unit's task organization, mission statement, Commander's intent, concept of operations, tasks to subordinate units, coordinating instructions, and control measures are key components of a plan. Commanders ensure their mission and end state nest with those of their higher headquarters. While the Commander's intent focuses on the end state, the concept of operations focuses on the way or sequence of actions by which the force will achieve the end state. The concept of operations expands on the mission statement and Commander's intent. Within the concept of operations, Commanders may establish objectives as intermediate goals toward achieving the operation's end state. When developing tasks for subordinate units, Commanders ensure that the purpose of each task nests with the accomplishment of another task, with the achievement of an objective, or directly to the attainment of an end state condition.

MISSION STATEMENT

The mission statement is a point of debate among planners. All planners agree the mission statement should include the who, what, when, where, and why of an operation. However, it is the what portion of the statement and the order in which they are organized that are the friction points. Many believe the what is the tactical mission task, and it is. Looking at the doctrinal definition of a tactical mission task A tactical mission task is the specific activity performed by a unit while executing a form of
tactical operation or form of maneuver (FM 3-90.1 and FM 6-0). The tactical mission task includes the action (essential task) and the form of maneuver or operation. FM 3-21-10 para 2-45 states the restated mission includes What (the unit's essential task and type of operation). Planners must understand this provides additional clarity for operations. For example, there are many types of raids and simply stating "conducts a raid" leaves room for confusion. Forces can conduct a raid to destroy, kill/capture or conduct a raid simply to secure an area or objective. Additionally, doctrine provides a list of actions forces can take. This list is simply a guide and does not require the use of just those action words. There is no definitive list of words or terms to describe the what and the why of a mission statement. The Commander is not limited to the tactical mission tasks listed in this appendix in specifying desired subordinate actions in an operations order or operations plan (FM 3-90.1). This is especially important for SOF operations since many of the SOF missions may not fit into this list. This is the same with the form of maneuver. SOF missions do not fall within the conventional forms of maneuver or operations. Instead, SOF planners should use SOF operations to fill in the type of operations. This is not to say SOF planners can't use the conventional operations. During a DA type mission, it would be appropriate to use the term raid. It is appropriate for SOF planners to use core mission as part of their mission statement (i.e. FID, UW, COIN). Commanders should use

tactical mission tasks or other doctrinally approved tasks contained in combined arms field manuals or mission training plans in mission statements (FM 6-0 para 9-68). This rule applies to the why, or the purpose, as well. For example, a SOF mission statement might look like this:

SFODA 91XX and PRF conducts <u>Unconventional Warfare</u> (type of op) operations to <u>coerce</u> (disrupt, or overthrow could also be used) (task) PRP and UPA forces within sector Gator NET DTG IOT <u>set conditions</u> (purpose) to delegitimize PRP and UPA forces.

The use of the word TO is the precursor for the tactical task, the term IOT is the precursor for the purpose.

The order in which the 5W's are organized should make sense. However, the who should be first, the what should always be type of operation then action, and the why (purpose) should always be last. If conducting operations with a partner force, planners should include them in the statement as well.

Common mistakes are to include the type of operation but not the tactical task and including tactical tasks within the purpose.

**NOTE- Examples given in the SF Detachment Mission planning guide are correct, however, the last sentence states "the op term could be removed and the statement would still have a clear task and purpose". The example given in FM 6-0 does not give an operational term either. However, if you read further it states "Commanders should use <u>tactical mission tasks</u> or other doctrinally approved tasks contained in combined arms field manuals or mission training plans in mission statements. These tasks have specific military definitions that differ from standard dictionary definitions. <u>A tactical mission task is a specific activity performed by a unit while executing a form of tactical</u>

operation or form of maneuver.". We recommend using a tactical mission task that consist of two parts: the form of maneuver or type of operation and the action verb (activity). This is correct according to what doctrine describes in the narrative.

On order missions should be included in the mission statement (usually after the complete mission statement). On order missions should include as much info as possible but no less than a what, where, and why. The mission statement may have more than one essential task (FM 6-0 para 9-66). Be prepared to missions should NOT be included in the mission statement. The difference between be prepared to and on order is the chance of occurrence. On order will happen, be prepared to might not.
When analyzing the higher order for specified and implied tasks, the staff also identifies any
be-prepared or on-order missions. **A be-prepared mission is a mission assigned to a unit that might be executed.** Generally, a contingency mission, Commanders execute it because something planned has or has not been successful. In planning priorities, Commanders plan a be-prepared mission after any on-order mission. **An on-order mission is a mission to be executed at an unspecified time.** A unit with an on-order mission is a committed force. Commanders envision task execution in the concept of operations; however, they may not know the exact time or place of execution. Subordinate Commanders develop plans and orders and allocate resources, task-organize, and position forces for execution (FM 6-0).

COMMANDER'S INTENT

The Commander's intent is essential to operations. Some would argue it is more important than the Mission Statement. It serves as the basis for why forces are conducting operations and why their operation is important to the overall mission, not just their mission. It also drives COA development. Commanders articulate the overall reason for the operation so forces understand why it is being conducted. They use the Commander's intent to explain the broader purpose of the operation beyond that of the mission statement. A well-crafted Commander's intent conveys a clear image of the operation's purpose, key tasks, and the desired outcome (ADRP 6-0). In order to keep a staff/ODA focused it is imperative that the XO/COS/180A understand the Commander's intent. The Commander's intent consists of three elements:

- Expanded Purpose
- Key Tasks
- End state

The expanded purpose- The expanded purpose provides Soldiers with a purpose of why their mission is essential to future operations or how their mission contributes to the overall mission. When describing the expanded purpose of the operations, the Commander's intent does not restate the "why" of the mission statement. Rather, it addresses the broader purpose of the operations and its relationship to the force as a whole (ADRP 5-0 para 2-94). It bridges the gap between higher mission and your mission.

EXAMPLE: the purpose of this operation is to delegitimize PRP Government & UPA forces within the sector by conducting ops by with and through PRF to win the popular support for PRF cause and force the withdrawal of UPA forces in sector to allow a successful transition and restore the legitimate ROP government.

Key Tasks- Key tasks are those tasks the unit, as a whole, needs to accomplish in order to accomplish the mission (ADRP 5-0). Commanders and planners should ask the question "if we fail this task can we complete the mission".

EXAMPLE:

- Proposed Key Task: Successfully infil undetected into AO.
  Ask the question: If the unit is compromised on infil do they fail the mission?
  If the answer is No, then it is not a good key task, if the answer is Yes then this is a good key task.
- Proposed Key Task: Link up with resistance
  Ask the question: If the element is not able to link up with the resistance, does the unit fail the mission?
  If the answer is Yes then this would be a good key task, if the answer is No then this would not be a good key task.

Planners should keep the number of key tasks to 3-5 in order to make it easy for Soldiers to remember and accomplish. Subordinates use key tasks to keep their efforts focused on achieving the desired end state (ADRP 5-0).

<u>Endstate-</u> The end state is a set of desired future conditions the Commander wants to exist when an operation is concluded (ADRP 5-0). It is described in in four sections:

- Friendly- to include host nation/partner
- Threat(s)
- Terrain
- Civilian

COA DEVELOPMENT

Preparation is key to course of action development as well as Commander's guidance. Commanders should have developed their COA evaluation criteria prior to COA Dev. Commanders should not disclose the criteria to the planners in order to ensure unbiased COA development. Planners should have access to all products produced prior to COA Dev such as higher's OPORD, and MA/IPB products (SITEMP MDCOA/MPCOA). Planners should be divided in Groups and there should be a common operating picture (COP) for all products (controlled by the XO/180A). The COP can be in the form of a PPT slide master or a master format for COA sketch/statement. This eliminates wasted time later in planning trying to organize all COAs into the same format.

During COA Dev there are several other critical actions that should take simultaneously by the staff (top four for ODA). This includes:

- 18A- Developing guidance for wargaming. The Commander should always be thinking 1-2 steps ahead and developing guidance.
- 180A- Running COA Dev by ensuring Commander's Intent is met and COAs are distinguishable
- 18Z- Ensuring timelines and troops to tasks are met
- 18F- Continue IPB (never stops)

All COAs must meet the same screening criteria:

- Feasible. The COA can accomplish the mission within the established time, space, and resource

limitations.

- Acceptable. The COA must balance cost and risk with the advantage gained.
- Suitable. The COA can accomplish the mission within the Commander's intent and planning

guidance.

- Distinguishable. Each COA must differ significantly from the others (such as scheme of

maneuver, lines of effort, phasing, use of the reserve, and task organization).

- Complete. A COA must incorporate—

How the decisive operation leads to mission accomplishment.

How shaping operations create and preserve conditions for success of the decisive operation or effort.

How sustaining operations enable shaping and decisive operations or efforts.

How to account for offensive, defensive, and stability or defense support of civil authorities' tasks.

Tasks to be performed and conditions to be achieved (FM 6-0).

The biggest friction point with the screening criteria is the distinguishability. In order to make a COA distinguishable from another, planners MUST look at the core mission and accomplish the mission differently. Simply changing the method of infil DOES NOT distinguish one COA from another.

COA Distinguishability Criteria

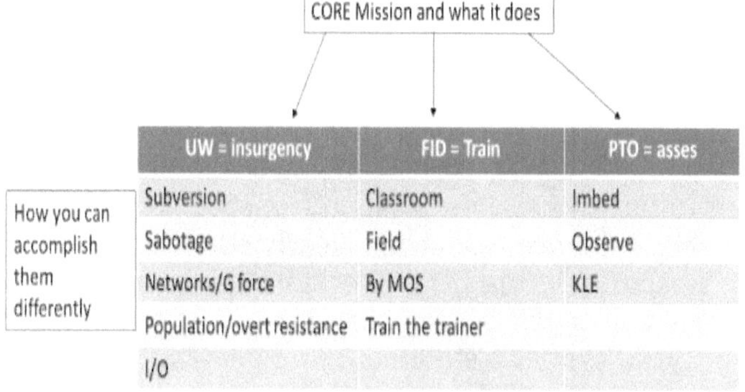

UW = insurgency	FID = Train	PTO = asses
Subversion	Classroom	Imbed
Sabotage	Field	Observe
Networks/G force	By MOS	KLE
Population/overt resistance	Train the trainer	
I/O		

UW Distinguishability Criteria

Population based (networks)	Vs.	Guerrilla based
Direct	Vs.	Indirect (surrogate)
Urban	Vs.	Rural
Kinetic (combat)	vs.	Non-Kinetic (I/O)

COA ANALYSIS (WARGAMING)

During Wargaming planners should think of the event as a sync or rehearsal. The XO/180A controls the event and there is an agenda and a time limit. During Wargaming:

- Must have the threat COAs (18F/S2 is the red hat)
- Planner do not have to wargame every COA
- Use graphics such as maps or models
- Use the wargaming to develop contingencies and decision points
- Record results

Problem Statement

The problem statement is another point of debate, particularly when it is done, what it accomplishes, and who writes it. First, it is important to understand that typically a problem statement comes from conducting Army Design Methodology which precedes MDMP. However, that should not preclude Commanders from conducting analysis and developing a problem statement as part of MDMP, it is a step in MDMP. After conducting analysis, the Commander should be able to develop his initial Commander's intent from the statement and issue planning guidance. Doctrine clearly states the problem statement can come before MDMP or during step 2 of MDMP (Conduct Mission Analysis- step 12 of MA). The advantage of writing an initial problem statement and presenting it to the staff early in planning, is that it helps focus the staff on MA efforts and becomes a part of the running estimate for the Commander. The Commander should write the statement. That is not to say he cannot receive assistance from his staff/team. When developing the problem statement, Commanders and planners must focus on what they can affect. Many times, Commanders will "peel the onion" to the point where they no longer can affect the problem. Additionally, it is not a reverse of the mission statement and it is a statement not a question. The best method for starting a problem statement is to begin with:

- "Keeping us from......."
- "Standing in the way of ...."

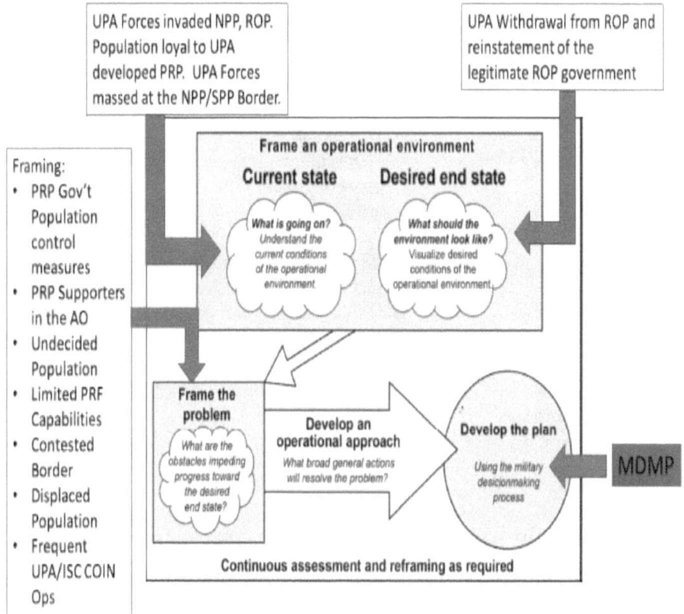

PROBLEM STATEMENT = Standing in the way of UPA forces withdrawing from ROP and the reinstatement of the legitimate ROP government are UPA forces massed on the NPP/SPP contested border, PRP government conducting population control measures along with COIN operations by the UPA and ISC, and an undetermined amount of population either supporting the UPA or are undecided.

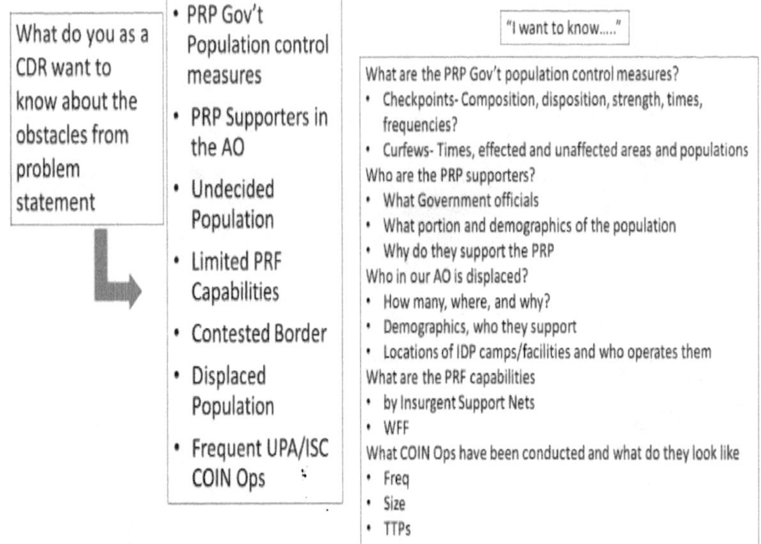

Leaders must truly look at the problem statement and realize that it may a very direct and uncomplex problem. For example, During the Normandy Invasion standing in the way securing Utah and Omaha Beaches and allowing Allied forces a foothold for the follow-on invasion forces, was a 100-foot cliff (Pointe du Hoc) with a battery of 155mm guns sitting on top.

MODIFYING THE MILITARY DECISION-MAKING PROCESS

It is important to note that Commanders and planners may modify the MDMP process based on time constraints. FM 6-0 para 9-12. The MDMP can be as detailed as time, resources, experience, and the situation permit. Performing all steps of the MDMP is detailed, deliberate, and time-consuming. Commanders use the full MDMP when they have enough planning time and staff support to thoroughly examine two or more COAs and develop a fully synchronized plan or order. This typically occurs when planning for an entirely new mission.

9-13. Commanders may alter the steps of the MDMP to fit time-constrained circumstances and produce a satisfactory plan. In time-constrained conditions, Commanders assess the situation, update the Commander's visualization, and direct the staff to perform the MDMP activities that support the required decisions. Here are a few examples of how the Commander might modify the MDMP process during time constrained planning:

- Commander issues guidance for only 2 COAs
- Commander integrates COA comparison into the COA Dev Brief, the Commander then selects the desired COA and the staff wargame only that COA, developing contingencies during wargaming.

Timelines and Actions

It is important for planner to use all time appropriately and there is generally some confusion on what actions to take after receiving the mission (in writing) or after receiving the Staff Mission Brief. Many times, planners are forced to wait on the Commander to issue his guidance for planning. A good practice for planners to take the following actions:

- 180A Develop the planning timeline (MA Brief, CCB, CBB, Syncs, IPRs)
- 18Z Assist with timeline, whitespace and develop tasks to accomplish. Issue a troop to task breakdown
- 18F Begin dissect the intel portion of the OPORD, divide and assign PMESII/ASCOPE
- Staff/ODA Begin developing running estimates from order consisting of
  -Facts
  -Assumptions
  -Specified tasks

-Implied tasks
-Essential tasks
-Determine constraints
-Review assets available
-Develop RFI/RFS from running estimates

- 18A- Develops initial guidance, Problem Statement.

Here are two examples of a timeline:

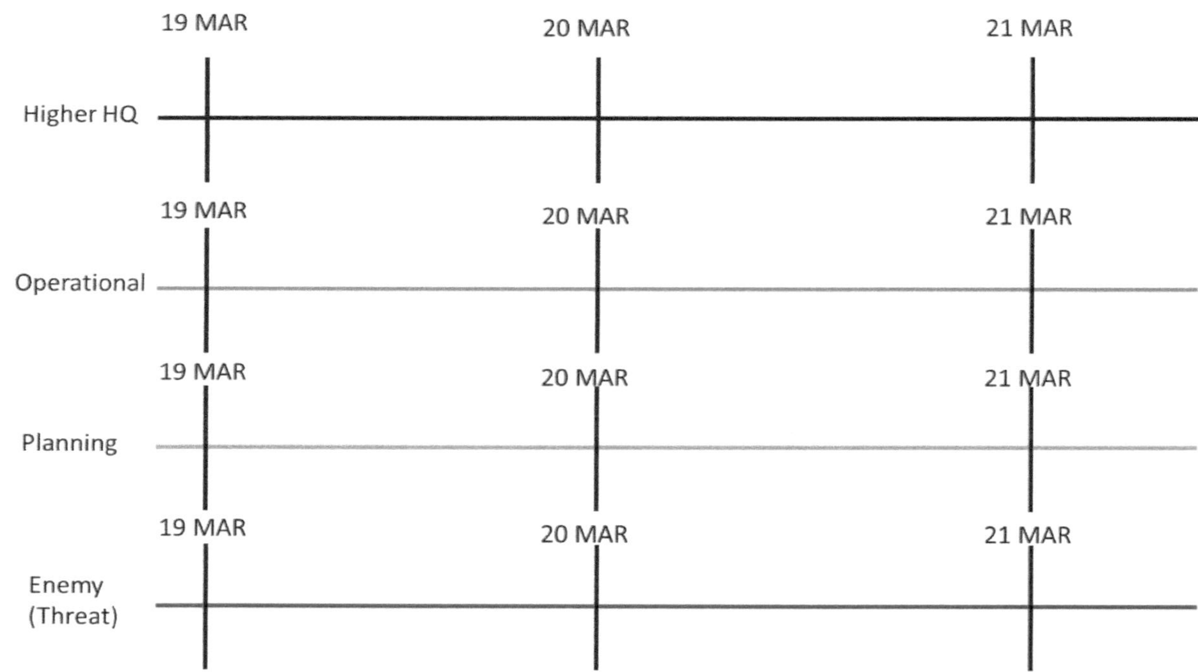
HOPE Timeline
19 MAR
20 MAR
21 MAR
Higher HQ
Operational
Planning
Enemy (Threat)

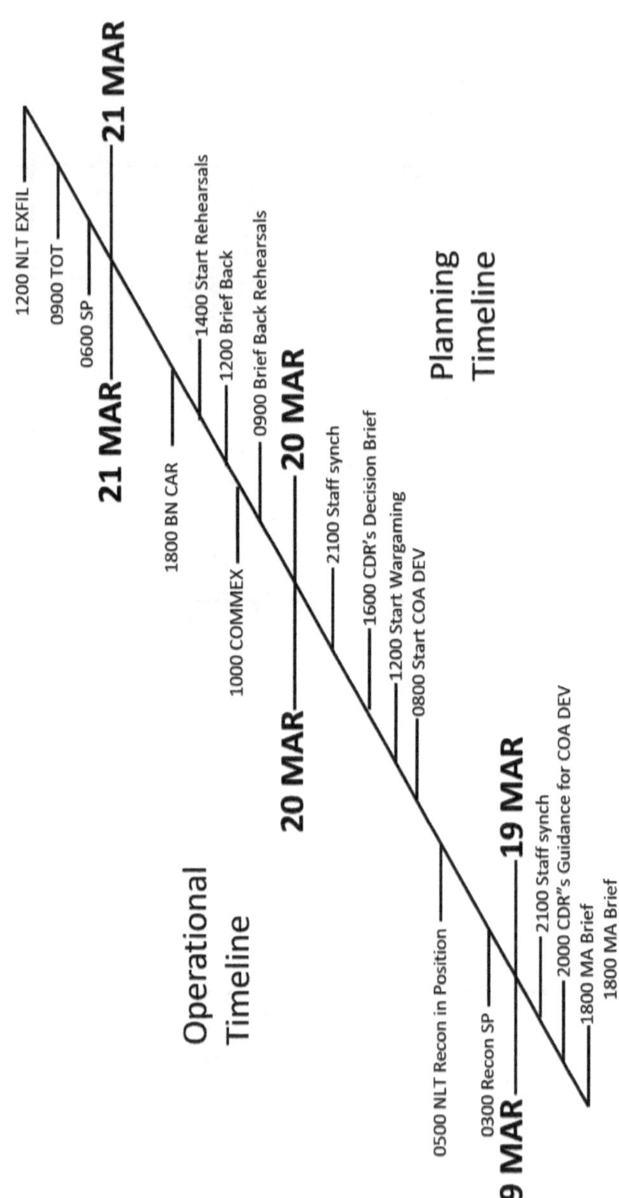
Operational
Timeline
19 MAR
0300 Recon SP
0500 NLT Recon in Position
20 MAR
1000 COMMEX
1800 BN CAR
21 MAR
0600 SP
0900 TOT
1200 NLT EXFIL
Planning
Timeline
1800 MA Brief
1800 MA Brief
2000 CDR"s Guidance for COA DEV
2100 Staff synch
19 MAR
0800 Start COA DEV
1200 Start Wargaming
1600 CDR's Decision Brief
2100 Staff synch
20 MAR
0900 Brief Back Rehearsals
1200 Brief Back
1400 Start Rehearsals
21 MAR

MOS Duties and Responsibilities for Planning

18B / AS3

- Fills the staff role of assistant S3
- Assists the 18Z with operations planning tasks; these may include but are not limited to:
  - Infil and Exfil plans
  - Route planning
  - Training plans
  - Fires planning
  - Air planning
  - Base defense and security plans
  - Risk management
  - Contingency planning
  - Developing assessments
  - ODA SOPs
  - Lethal targeting
  - ISR synchronization with 18F
  - Briefing parts of Scheme of Maneuver
- Assists 180A with EPA plan
- Assists 18F with IPB (OAKOC and PMESII analysis)
- Advises ODA on weapon employment
- Responsible for the function, maintenance, and security of team weapons
- Responsible for requesting, securing, handling, and issuing of ammo and pyro
- Cross trains ODA in weapons tasks

18C / S4

- Fills the staff role of S4
- Responsible for all logistical considerations for the ODA
  - Forecasting classes of supply and sending up requests based on shortages
  - Procurement methods when deployed
  - Cache planning
  - Resupply planning
  - LZ/DZ operations
  - Rigging bundles
- Responsible for developing and briefing the sustainment plan for the CCB and the CBB
- Advises ODA on Engineer operations
- Provides in depth targeting analysis utilizing CARVER
- Assists 18F with IPB (targeting and PMESII analysis)
- Develops load/cross load plan
- Plans for the development of logistical/sustainment networks for both the ODA and partner forces
- Maintains property accountability records for all team equipment
- Funds management; oftentimes a Pay Agent (PA) or Field Ordering Officer (FOO)
- Cross trains the ODA in engineer tasks

18D / S1

- Fills the staff role of S1
- Responsible for developing and briefing the medical evacuation plan for the CCB and the CBB
- Forecasts and requests medical supplies needed for both the ODA and the partner force
- Fulfills the administrative functions required for the ODA such as:
  - Tracking messages, RFIs, RFSs, and products that go in and out of the ISOFAC
  - Preparing manifests
  - Submitting PERSTATs as required by Higher HQ
  - Establishing and posting the access roster for the ISOFAC
  - Ensuring all ODA medical records are updated
  - Conducting daily sick call for the ODA and partner force
  - Recorder keeper for the partner force (pay, awards, rosters, training conducted, medical status)
- Advises the ODA on medical considerations for the mission
- Assists the 18F with IPB (PMESII analysis)
- Assists in the development of medical networks and infrastructure for the partner force
- Cross trains the ODA in medical tasks

18E / S6

- Fills the staff role of S6
- Responsible for the development and briefing of the Communications plan in the CCB and CBB; may include:
  - PACE plans for the ODA internal, to higher HQ, and to the partner force
  - Brevity codes
  - OPSKEDs
  - Challenge and Passwords, Running Passwords, Number combinations
- Responsible for communication and cryptographic equipment requests, accountability, and maintenance
- Assists the partner force with development of communications networks and training
- Responsible for the requesting of frequencies
- Plans for the distribution and cross loading of communications equipment
- Conducts PCCs/PCIs on all communications equipment prior to SP
- Advises the ODA on communication considerations for the mission
- Cross trains the ODA in communication tasks

Duties and Responsibilities for Planning (Big Four)

18A CDR "Mission"

- Responsible for everything the team does or fails to do
- Fills the staff role of CDR during planning
- Uses mission command to direct the operations cycle; exercises the six principles of mission command
- Thinking at least 1-2 steps ahead of the staff during MDMP
- Issues clear and concise guidance to the staff at critical points of the planning process
- Continually assesses MDMP process to ensure it stays within the mission guidance
- Responsible for **managing "the mission"**; ensures all ODA members have an accurate understanding of the mission
- Develops a clear commander's intent, mission statement, and problem statement (should use key staff members to provide input and feedback)
- Face of the ODA and advisor to partner, host nation, or insurgent leadership

180A XO "Systems"

- Fills the staff role of XO during planning
- Responsible for ensuring CDRs guidance is adhered to during planning
- Chief of staff for the other ODA MOSs
- Develops and **manages "the systems"** the ODA utilizes during planning, these may include but are not limited to:
    - RFI/RFS format and trackers
    - Managing IPRs and Staff Synchs
    - QAQC all planning products
    - ISOFAC layout and procedures
    - Digital vs analog MDMP products
    - Systems for TOC operations and battle tracking
    - Formats for briefings
- Responsible for the EPA
- Responsible for running briefing rehearsals for the ODA
- ASOT Manager for ODA
- CDR for 2nd split team
- CDR in absence of 18A
- Develops the MDMP planning timeline
- Manages the Wargame during COA Analysis
- A primary advisor to the CDR

18Z S3 "Men"

- A primary advisor to the CDR
- Fills the staff role of S3 during planning
- Develops and briefs the scheme of maneuver for the Commander's Brief Back with assistance from other staff members
- Responsible for development of the training plans; this can include training plans for:
    - ODA cross training
    - ODA SOPs
    - Partner force training
- Responsible for **managing "the men"**; uses tools to maintain accountability and enforce standards such as:
    - Keeper and enforcer of the timeline; populates events with input from the 180A
    - Develops, maintains, and enforces a troops to task matrix to keep the staff on track
- Responsible for tactical rehearsals, PCIs/PCCs, and maintenance
- Responsible for the day to day Battle Rhythm of the ODA

18F S2 "Intel"

- Fills the role of the S2 during planning
- Responsible for **managing "intelligence"** which includes at a minimum:
    - Provide guidance to staff members assisting, and ensures that all four steps of IPB are complete and backed with good analysis
    - Roles up IPB products into Threat COAs
    - Maps threat networks
    - Primary briefer for IPB products
    - Identifies information gaps, develops PIR, and builds the Information Collection Plan
- Advises CDR on intel processing, targeting, site exploitation, and tactical questioning
- Provides guidance to assist the 180A in ASO operations
- Informs the 18Z when manpower requirements need adjustment
- Continually updates the operating picture (threats and friendly)
- Responsible for classified info

Index

Table of Figures

Glossary

1st SFC(A) – 1st Special Forces Command (Airborne)
AAR – After Actions Review
ACFT – Army Combat Fitness Test
ADVON – Advanced echelon
AGCM – Army Good Conduct Medal
AHA – Ammunition Holding Area
AI – Assistant Instructor
AO – Area of Operations
AOB – Advanced Operations Base
AOR – Area of Responsibility
APFT – Army Physical Fitness Test
AR – Army Regulation
ARMY PUBS – Army Publishing Directorate
ARSOF – Army Special Operations Forces
ASCOPE – Areas, Structures, Capabilities, Organizations, People, Events
ASI – Additional Skill Identifier
ASP – Ammunition Supply Point
ATP – Army Techniques and Procedures
ATRRS – Army Training Requirements and Resources System
AXP – Ambulance exchange Point
BASD – Basic Active Service Date
BMQ – Basic Military Qualification (found on SRB, for Rifle qualification)
BN – Battalion
BSM – Bronze Star Medal
CANSOF – Canadian Special Operations Forces
CARVER – Criticality, Accessibility, Recuperability, Vulnerability, Effect, and Recognizability
CATS – Combined Arms Training Strategies
CBB – Commanders Back Brief
CCB – Commanders Concept Brief
CDQC – Combat Diver Qualification Course
CG – Commanding General
CI – Chief Instructor
C-IED – Counter Improvised Explosive Device
CNT – Counter-Narcoterrorism
CO – Commanding Officer
Co – Company

COA – Course of Action
COA DEV – Course of Action Development
COIN – Counter Insurgency
COL – Colonel
CONOP – Concept of Operation
CONUS – Continental United States
COP – Common Operating Picture
COR – Change of Rater (NCOER)
CPAP – Continuous Positive Airway Pressure
CSM – Command Sergeant Major
CTC – Combat Training Center
CTR – Compete the Record (NCOER)
DA – Department of the Army
DA PAM – Department of the Army Pamphlet
DD – Department of Defense
DIRLAUTH – Direct Liaison Authorized
DLPT – Defense Language Proficiency Test
DOR – Date of rank
DTMS – Digital Training Management System
DTS – Defense Travel System
DZ – Drop Zone
E&E – Escape and Evasion
EDRE – Emergency Deployment Readiness Exercise
EFMB – Expert Field Medical Badge
EIB – Expert Infantryman's Badge
ELO – Enabling Learning Objective
ETS – Expiration-Term of Service
FID – Foreign Internal Defense
FIDEX – Foreign Internal Defense Exercise
FMP – Full Mission Profile
FNG – Fucking New Guy
FOOM – Formations, Order of Movement
FRAGO – Fragmentary Order
FSC – Forward Support Company
FWD – Forward (deployed)
GAF – Ground Assault Force
GCC – Geographical Combatant Commander
GED – General Education Diploma

GOTWA – Where is the Leader **Going**, **Others** taken with the leader, **Time** the leader will be gone, **what** to do if the leader does not return on time, **Actions** taken upon contact
GOVCC – Government credit card
GPOT – Graphic Depiction of Terrain
GRP – Group
GSB – Group Support Battalion
GTA – Graphic Training Aid
HAF – Helicopter Assault Force
HAZMAT – Hazardous Materials
HHC – Headquarters, and Headquarters Company
HLZ – Helicopter Landing Zone
HN – Host Nation
HPTL – High Payoff Target List
HQ – Highly Qualified (NCOERs)
HRC – Human Resources Command
HSC – Headquarters Service Company
HVT – High Value Target
HVTL – High Value Target List
IAW – In Accordance With
IDF – Indirect Fire
IED – Improvised Explosive Device
IFAK – Individual First Aid Kit
IOT – In Order To
IPB – Intelligence Preparation of the Battlefield
IPR – In-Process Review (informal brief)
ISOFAC – Isolation Facility
JCET – Joint Combined Exercise Training
JM – Jumpmaster
JMRC – Joint Multinational Readiness Center
JPAT – Joint Planning and Assistance Team
JRTC – Joint Readiness Training Center
JSOFSEA – Joint Special Operations Forces Senior Enlisted Academy
JSOU – Joint Special Operations University
JTAC – Joint Terminal Attack Controller
km – Kilometer
LNO – Liaison Officer
LPD – Leader Professional Development
LRTC – Long Range Training Calendar

LTC – Lieutenant Colonel
LTG – Lieutenant General
MA – Mission Analysis
MASCAL – Mass Casualty (event)
MCOO – Modified Combined Obstacle Overlay
MDCOA – Most Dangerous (deadly) Course of Action
MDMP – Military Decision Making Process
MEDEVAC – Medical Evacuation
MEL – Military Education Level
MES – Military Education Status
MET – Mission Essential Tasks
MFF – Military Free Fall
MFFJM – Military Free Fall Jumpmaster
MILPER – Military Personnel (official message traffic from HRC)
MLC – Master Leader Course
MLCOA – Most Likely Course of Action
MOS – Military Occupational Specialty
MP – Military Police
MQ – Most Qualified (NCOERs)
MRC – Medical Readiness Code
MSG – Master Sergeant
MSM – Meritorious Service Medal
MSS – Mission Support Site
MTOE – Modification Table of Organization and Equipment
NCO – Non Commissioned Officer
NCOER – Non Commissioned Officer Evaluation Report
NCOES – Non Commissioned Officer Education System
NCOIC – Non Commissioned Officer In Charge
NCOPD – Non Commissioned Officer Professional Development
NET – No Earlier Than
NLT – No Later Than
NODS – Night Optic Devices
NTC – National Training Center
NVG – Night Vision Goggle
OAKOC – **Observation** and Fields of Fire, **Avenues** of Approach, **Key** Terrain, **Obstacles**, **Cover** and Concealment
OCONUS – Outside the Continental United States
OIF – Operation Iraqi Freedom
OML – Order of Merit List

OPCEN – Operations Center
OPI – Oral Proficiency Interview
OPLAN – Operation Plan
OPORD – Operations Order
OPSKED – Operational Schedule
ORP – Objective Rally Point
OS – Overseas
OT – Occupational Therapist
PCC – Pre Combat Checks
PCI – Pre Combat Inspections
PCS – Permanent Change of Station
PDSS – Pre Deployment Site Survey
PF – Partner Force
PI – Primary Instructor
PLANEX – Planning Exercise
PMCS – Preventive Maintenance Checks and Services
PME – Professional Military Education
PMESII – Political, Military, Economic, Social, Information, and Infrastructure
PMOS – Primary Military Occupational Specialty
PMT – Pre Mission Training
POI – Program of Instruction
PT – Physical Therapist
PTSD – Post Traumatic Stress Disorder
PX – Post Exchange
QTB – Quarterly Training Brief
RFF – Request For Forces
RFI – Request For Information
RFS – Request For Support/Supplies
RMT – Realistic Military Training
RON – Remain Overnight
RSM – Resolute Support Mission (2015-2021)
RSO – Range Safety Officer
RTB – Ranger Training Brigade, and Return to base
S1 – Personnel actions
S2 – Intelligence
S3 – Operations
S4 – Logistics
S5 – Future plans, not all BNs maintain this shop

S6 - Communications
SF – Special Forces
SFARTAETC – Special Forces Advance Reconnaissance, Target Analysis, and Exploitation Techniques Course
SFAS – Special Forces Assessment and Selection
SFAUC – Special Forces Advanced Urban Combat
SFC – Sergeant First Class
SFG – Special Forces Group
SFISC – Special Forces Intelligence Sergeants Course
SFOD-A – Special Forces Operational Detachment-Alpha
SFOD-B – Special Forces Operational Detachment-Bravo (or written as: B-TEAM, AOB)
SFPDM – Special Forces Professional Development Model
SFQC – Special Forces Qualification Course
SFSC – Special Forces Sniper Course (formally known as Special Operations Target Interdiction Course SOTIC)
SGLV – Servicemembers' Group Life Insurance Election and Certificate
SGM – Sergeant Major
SITREP – Situation Report
SLC – Senior Leaders Course
SLC – Senior Leaders Course
SLJM – Static Line Jumpmaster
SM – Service Member
SMOS – Secondary Military Occupational Specialty
SOCMSSC – Special Operation Combat Medic Skills Sustainment Course
SOF – Special Operations Forces
SOP – Standard Operating Procedure
SOTAC – Special Operations Terminal Attack Controller
SOTF – Special Operations Task Force
SPC – Specialist
SQI – Special Qualification Identifier
SR – Senior Rater
SR – Special Reconnaissance
SRB – Soldier Record Brief
SRP – Soldier Readiness Processing
SSG – Staff Sergeant
STP – Soldiers Training Publication (MOS REG)
STRAC – Standards in Training Commission
SUT – Small Unit Tactics
SVEST – Suicide Vest (IED)

SWCS – Special Warfare Center and School (shorthand for United States Army John F. Kennedy Special Warfare Center and School USAJFKSWCS)
SWTC – Special Warfare Training Course
SWTG (A) – Special Warfare Training Group (Airborne)
TASKORG – Task Organization
TC – Training Circular
TCCC – Tactical Combat Casualty Care
TDA – Table of Distribution and Allowances
TDY – Temporary Duty
THOR3 – Tactical Human Optimization Rapid Rehabilitation and Reconditioning
TLO – Terminal Learning Objective
TLP – Troop Leading Procedures
TOC – Tactical Operation Center
TRP –Target Reference Point
TS –Top Secret
TSCP – Theater Security Cooperation Program
TSOC – Theater Special Operations Command
TST – Time Sensitive Target
TTP – Tactics, Techniques, and Procedures
UMO – Unit Movement Officer
USASOC – United States Army Special Operations Command
USSOCOM – United States Special Operations Command
UW – Unconventional Warfare
VBIED – Vehicle Born Improvised Explosive Device
VTAC – Company owned by Kyle Lamb
VW – Voluntary Withdrawal
VWCPT – Visibility, Winds, Cloud Cover, Precipitation, Temperature/humidity
WARNO – Warning Order

A poem that struck home years after my team time. I hope it speaks to you as it did to me.

Is anybody happier because you passed his way?
Does anyone remember that you spoke to him today?
This day is almost over, and its toiling time is through;
Is there anyone to utter now a kindly word of you?

Did you give a cheerful greeting to the friend who came along?
Or a churlish sort of "Howdy" and then vanish in the throng?
Were you selfish pure and simple as you rushed along the way, or is someone mighty grateful for a deed you did today?

Can you say tonight, in parting with the day that's slipping fast, That you helped a single brother of the many that you passed?
Is a single heart rejoicing over what you did or said;
Does a man whose hopes were fading now with courage look ahead?

Did you waste the day, or lose it, was it well or sorely spent?

Did you leave a trail of kindness or a scar of discontent?
As you close your eyes in slumber do you think that God would say, you have earned one more tomorrow by the work you did today? – Edgar Guest

Acknowledgements

I would like to thank the following individuals for their contributions to this book. I knew going into this project that I did not have all the answers after my time in the seat. Had all of the following guys not shared their experience and knowledge this book would have never seen the light of day.

MSG Kenneth Powell (0322), thank you for asking me about my Team Sergeant time. Without it, I would have never started this project. Your questions during our talk served as the catalyst for my honest self-reflection.

1SG (RET) Enrique Longoria (5135), your contribution provided legitimacy to the information in this book. To have a successful Team Sergeant from a different Group, corroborate my work has been invaluable.

MSG (RET) Aaron Marlow (5312), Thank you for entertaining all of my rants when it came to the topics in this book, or anything for that matter. More importantly, had it not been for the "Marlow product," Chapter 3 would have been a disorganized mess.

SGM (RET) Adam Delgado (5225), Thank you for your honest feedback. It was the push I needed to make this book to what it is today.

SGM (RET) Bruce Schnabel (9526), thank you for the countless hours discussing the topics in this book, and the hundreds of rabbit holes we dove into. If I remember correctly, you were the first person who suggested to write this book.

SGM Jerry Wilson (9525), your candid feedback focused my efforts to provide the needed details for an up-and-coming Team Sergeant.

MSG Matthew Smith (0436), you have always been the voice of reason in all matters concerning the ODA. Thank you for keeping me grounded, which in turn kept the book usable for all ODAs.

MSG (RET) Paul LeFavor (3233), Thank you for accepting my book into your publishing Company. Also, many thanks for the hours spent discussing what needs to be said to reach our future leaders.

CSM (RET) Rob Flournoy (091), your presence, influence and mentorship were bigger than life in the hall of C Co, 3rd BN. Even now, you are still educating me on the finer workings of strategic level decision making. Thank you for taking the time to read my book and write the Foreword.

CSM (RET) Bill Hanes (0136), thank you for expert tutelage on all things admin. More important than the information you shared; you explained WHY and how the administrative tasks run the system.

CSM Kevin Dorsh (ODA 0414), thank you for volunteering your time to serve as the editor, your strong endorsement of this project, and for the countless hours spent discussing all of the topics in the book.

SFC Brian Heft, within a short period of working for you I knew what kind of leader I wanted to be when I grew up. Granted we met back in my Infantry days, but you were the first NCO I worked for who's primary purpose was to exceed every standard. Your drive pushed the squad further than we thought we could go. Thank you for your absolute dedication to training, coaching and mentoring. Your example became my foundation for this book.

About the Author

Master Sergeant Thomas Kelly (RET) served 24 years on active duty 20 of which within Special Forces. Prior to the events on 9/11, he was an Airborne Infantryman assigned to Apache Company, 1/501st PIR, Fort Richardson, AK. Post 9/11, after being selected, and completing the SFQC, he served as an Engineering Sergeant, Intelligence Sergeant, and as a Team Sergeant in 10th Special Forces Group. Post team time MSG Kelly served as the Chief Instructor at the SOF Mountaineering school, at 5/19 National Guard SF as a LNO, and finally as a Chief Instructor for the SUT phase of the SFQC. Notable deployments: Kosovo, Iraq-OIF III, V, VI, South Africa, Azerbaijanian, Niger, Jordan, and Afghanistan-RSM 11.

Team Sergeant, SFOD-A 0332 SEP 2009 – DEC 2012.

Connect with Blacksmith Publishing

www.thepinelander.com

www.blacksmithpublishing.com

www.ingramcontent.com/pod-product-compliance
Lightning Source LLC
Chambersburg PA
CBHW021701120626
46545CB00004B/1348

* 9 7 8 1 9 5 6 9 0 4 1 1 6 *